Vancouver Is Ashes

The Great Fire of 1886

Lisa Anne Smith

RONSDALE PRESS

RONSDALE PRESS
3350 West 21st Avenue, Vancouver, B.C. Canada V6S 1G7
www.ronsdalepress.com

Typesetting: Julie Cochrane, in Granjon 11.5 pt on 16
Cover Design: Shed Simas
Cover Photo: Vancouver Tent City Hall (VPL 1089A). Photo, Harry T. Devine.
Paper: Ancient Forest Friendly Rolland 60 lb. Opaque, FSC Recycled,
100% post-consumer waste, totally chlorine-free and acid-free.

Ronsdale Press wishes to thank the following for their support of its publishing program: the Canada Council for the Arts, the Government of Canada through the Canada Book Fund, the British Columbia Arts Council, and the Province of British Columbia through the British Columbia Book Publishing Tax Credit program.

Library and Archives Canada Cataloguing in Publication

Smith, Lisa, 1959–, author
Vancouver is ashes: the great fire of 1886 / Lisa Anne Smith.

Includes bibliographical references and index.
Issued in print and electronic formats.
ISBN 978-1-55380-320-1 (print)
ISBN 978-1-55380-322-5 (ebook) / ISBN 978-1-55380-321-8 (pdf)

1. Vancouver (B.C.)—History—19th century. 2. Vancouver (B.C.)—History—19th century—Sources. 3. Fires—British Columbia—Vancouver—History—19th century. 4. Fires—British Columbia—Vancouver—History—19th century—Sources. I. Title.

FC3847.4.S579 2014 971.1'3303 C2014-900053-7 C2014-900054-5

At Ronsdale Press we are committed to protecting the environment. To this end we are working with Canopy (formerly Markets Initiative) and printers to phase out our use of paper produced from ancient forests. This book is one step towards that goal.

Printed in Canada by Marquis Book Printing, Quebec

for my mother
Eileen Boberg,
who shares my passion
for history

CONTENTS

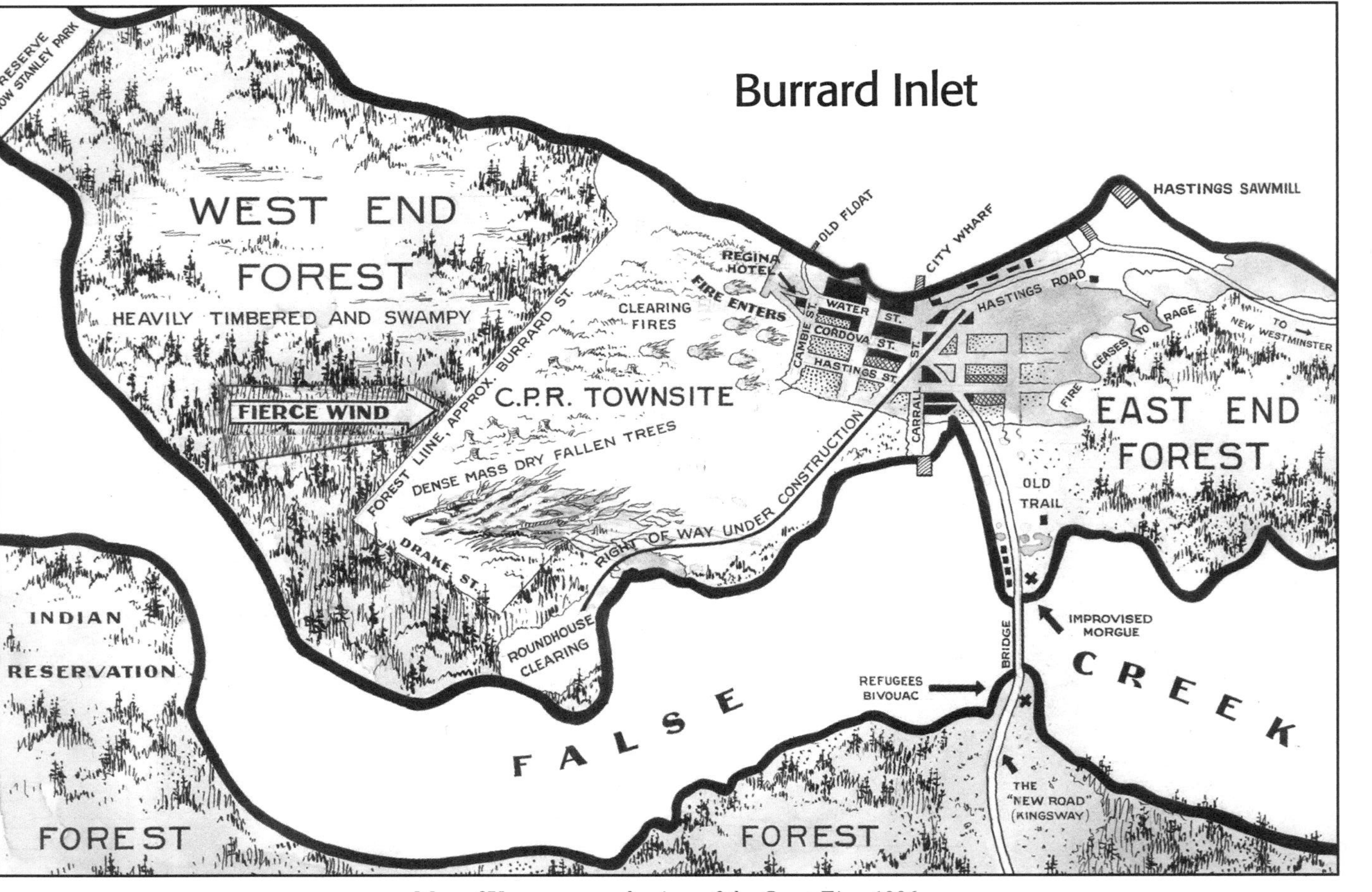

Map of Vancouver at the time of the Great Fire, 1886.

Tuesday, June 8, 1886

Vancouver is one of the liveliest cities on the Pacific Coast. With its wonderful natural advantages, its glorious climate, enchanting scenery, its unrivalled harbour, no wonder the prediction is freely made that in a few years there will be seen thousands of people where there are now hundreds.

Vancouver Daily Advertiser

Sunday, June 13, 1886

■ 10 A.M.

Vancouver is sweltering. Late spring has been abnormally hot for the past three weeks. Puddles, normally knee-deep along busy Carrall Street, have long since dried up. The few remaining mud holes are blistered and cracked like paint on an ancient canvas. Horses swish their tails in a half-hearted attempt to keep the ever-increasing population of flies at bay. Chickens scratch up small clouds of dust as they scavenge about their backyard enclosures.

Reverend Thompson's deep voice echoes from the rafters of Vancouver's recently completed Presbyterian Church, as he reads from the Bible laid open upon his small pulpit: "And suddenly there came a sound from heaven as of a rushing mighty wind, and it filled all the house where they were sitting. And there appeared unto them cloven tongues like as of fire, and it sat upon each of them."

It is Whitsunday, exactly fifty days after Easter, and the morning's sermon is all about the miraculous events described in the second chapter of "The Acts of the Apostles." Reverend Thompson artfully captivates his audience. Parishioners sit transfixed, fanning themselves in the morning heat. The normally full church is emptier than usual today. Many members of the congregation have travelled to New Westminster for the funeral of Canadian Pacific Railway contractor Owen McCormack, killed on the job when the dynamiting of a stump went horribly wrong. Vancouver, newly incorporated city as it is, does not yet have its own official cemetery. Aboard a small wagon, the body of Mr. McCormack had to be transported through twelve miles of near-virgin wilderness for internment in accordance with provincial law.

Despite the lack of burial facilities, Vancouver is well on its way to becoming "civilized." In the previous year, CPR General Manager William Van Horne had confirmed that the small collection of buildings formerly known as "Granville" was to become the western terminus of the Canadian Pacific Railway, rather than the Burrard Inlet headwater township of Port Moody. Now, the pace of development is surging forward. New businesses are going up as fast as land can be purchased. Stages and steamers discharge luggage-laden settlers, who find lodging in hotels still smelling of freshly milled cedar. Speculators ponder survey maps carefully spread over rough-hewn tables, while raucous lumberjacks drink away their earnings in smoke-filled saloons along the length and breadth of Water Street.

From the earliest days of settlement on Burrard Inlet's south

At the corner of Carrall and Water streets, May 1886.

shore, trees have been the drawing card. The indigenous Musqueam, Squamish and Tseil-Waututh peoples lived resourcefully off the land, selectively harvesting the verdant growth of the region for everything from sturdy longhouses and dugout canoes to boxes and baskets of intricate beauty. When an enterprising Captain Edward Stamp established the B.C. and Vancouver Island Spar, Lumber and Sawmill Company on the natural promontory near the seasonal native community of Kumkumalay (big leaf maple trees), trees meant money in the pocket and food on the table.[1] The abundant western red cedars have trunks so wide lumberjacks must mount springboards to reach sections narrow enough to saw through. Douglas firs tower skyward, some reaching three hundred feet or more in height. They are the ideal length for use as sailing-ship masts and spars.

For over twenty years, the steady rasp of the crosscut saw has cut a wide swath into the thick stands of ancient-growth forest flanking the inlet. An ingenious system has been developed for speeding up the pace of the work. The tallest of the trees are sawed strategically to knock down partially-axed surrounding trees as they fall, like a giant line of dominoes. Try anything, but "get it down," is the prevailing mindset.[2]

After being stripped of all branches, the massive trunks are skidded to the mills over a corduroy road of logs laid side-by-side and greased with dogfish oil. The leftover heaps of slash and debris are burned away, the stumps dynamited, and then the land can be graded for construction to begin. Felling, slashing and burning continue relentlessly from dawn to dusk. At the end of each long day, the powder men apply their torches to the fuses and great, hulking stumps shatter—their deeply embedded roots rocketing skyward. The speed of felling far outpaces the speed of burning, clear up and stumping. A virtual mountain of slash, up to three storeys high in places, has been steadily accumulating on the outskirts of the city. Vancouver residents have long become accustomed to the sounds of industry and the acrid smell of smoke lingering in their nostrils. Such is the price of progress.

Earliest known photo of Hastings Sawmill and townsite, 1872.

Plans have been drawn up for the construction of a roundhouse near the southerly end of the future Drake Street to turn the CPR locomotives about-face for their return journey eastward. The branch line to False Creek will provide freight service for the Royal City Planing Mills and other industries soon to occupy the English Bay foreshore. A work party keeps vigil over a small clearing fire at the roundhouse site. Efforts are low-key, partly due to it being Sunday and partly due to the relentless heat—already taking its toll on motivation. Far out on the broad expanse of English Bay, a gentle Pacific breeze wafts its way eastward, causing the previously calm waters of False Creek to slosh rhythmically against the shoreline. The crewmen keep buckets filled with water on standby, ready to deal with the sudden flare-ups which inevitably occur in dry weather.

William John Gallagher

Near the future intersection of Cambie and Smithe streets, another company of workers steadily hack and shovel through the profuse undergrowth of salal and salmon berry. They are hired by "Percival and Gallagher,"[3] the company contracted by the Townsite Commission to clear a branch line from the CPR terminus of Vancouver centre, to False Creek. Ontario-born William Gallagher, already well on his way to an enterprising business career at the age of twenty-three, has charge over the rail-bed crew. He strides back and forth, energetically monitoring progress to ensure that every last vestige of root is dug out, well aware that the notoriously prolific salmonberry can fight its way back through the thickest layer of gravel. When complete, the rail bed will cut a northeast trajectory across the mile-wide peninsula between False Creek and Burrard Inlet, linking with the main line.

With his own crew working diligently, Gallagher decides to wander down and have a look at the roundhouse clearing site. He quickly discovers that the men have a struggle on their hands. Small flames stirred to life by the offshore breeze are licking hungrily at the ample supply of deadwood. Showers of sparks teasingly dance in the air each time a shovel is rammed down. Wet blankets need fast and frequent re-soaking at the False Creek shore. Somewhat perturbed, Gallagher hurries back to his worksite to ask for volunteers to help. Flare-ups are common at the clearing sites and the unspoken law among crewmen is to provide aid wherever and whenever it is needed. Three men immediately offer their assistance.

■ 12 NOON

The Presbyterian Church congregation has filtered out into the early afternoon sunshine. Small groups of parishioners linger to exchange pleasantries—ladies clandestinely drawing deep breaths of air to offset the discomforts of their tightly-corseted Sunday dresses, business-suited gentlemen idly puffing on cigars. Fifteen-year-old Annie Ellen Sanders tries to keep a wary eye on her younger brothers Joseph and James, while chatting with a group of young men in the crowd. She does not feel particularly flirtatious. Her elder sister Catherine died less than two months previously and the Sanders family are, once again in mourning. Four sisters have died over the past few years and Annie Ellen, last surviving female member of the Sanders children, has found herself thrust into the role of guardian and keeper of her brothers. She listens with only mild interest while four young men discuss their plans for the day.

"This is a grand day to burn those branches and bushes!" one of them remarks.

"Well, if you do it, we will too,"[4] another agrees.

Two of the young men are helping to clear land on Granville Street, others doing the same at Hastings and Cambie streets. The fact that it is the Sabbath cannot hold back their youthful enthusiasm. "So long!" they call to Annie Ellen, as they troop off to their various projects. The Sanders family head for their Prior Street home on the eastern outskirts of Vancouver to have lunch, after which it will be time for Annie and her brothers to return to the church for Sunday-school class.

Lunch breaks at the roundhouse clearing site are taken in turn, as efforts to control the fire continue. The men are on the run to and from the water's edge, scooping up bucketfuls of water and saturating every available blanket. William Gallagher has remained to monitor the situation and sees that it is steadily worsening. The stubborn offshore breeze has refused to abate and the flames are growing in height and ferocity. Just how far the danger may spread is questionable, and he soon begins to worry about the Percival and Gallagher office at the foot of Columbia Street, where wages and accounting records are stored under lock and key. Before he leaves, Gallagher issues a stern warning to the crew. "If the fire breaks away from the clearing, do not attempt to fight it, or you will lose your lives!"[5]

Sunday, June 13, 1886

■ 12:30 P.M.

Midday dinner on Sunday is typically a social affair in Vancouver. In the elite residences on the eastern side of the city, servants are hard at work throughout the morning preparing the meal while their employers attend church. From the cavernous interiors of wood-burning stoves, a mouth-watering array of courses emerges—new spring potatoes nestled alongside a roast of beef dripping with gravy, freshly caught salmon, duck with currant jelly and croquettes of rice—all to be placed upon oak dining tables spread with fine Irish linen, silverware polished to a glossy sheen and the most extravagant of English bone china. Other households tuck into simpler fare—slices of homemade bread with spring lettuce or home-cooked strawberry conserve, slices of cheese—all washed down with a cup of hot tea or coffee from the kettle permanently occupying the stove back burner. Dinner may be followed by an afternoon stroll in fine weather, or a much-coveted nap, purely to rest weary bones from a week's worth of labour.

The crewmen continue to work feverishly at the roundhouse site, bolstered by additional hands from Percival and Gallagher. Buckets of water and shovels are fast becoming inadequate. Any attempt to clear a firebreak through the enormous tangle of deadfall would be impossible. It would require days of work. Oxen, needed to haul away the logs, would be roasted alive. Defeated by the flames and with no other safe direction to turn, the crewmen are forced to abandon their efforts and retreat to the False Creek shoreline. Out on the water, natives from the south shore village of Snauq are paddling across in a flotilla of canoes, alarmed that the typically grey haze to the north has suddenly erupted into orange.

■ 1 P.M.

Creatures of the forest and rangeland have a built-in survival mechanism. When fire threatens, they instantly deploy a variety of defensive behaviors. As flames progress up the gentle slope from False Creek, birds instinctively take to the skies, deer and coyotes run, rodents, frogs and salamanders burrow deep underground. The preceding months of habitat disturbance and destruction have already caused many of Vancouver's forest dwellers to abandon their accustomed homes. The remains of eagle nests are scattered amidst the deadfall. Owls and herons have long departed in search of more idyllic surroundings. Black bears, once regular habitués of the inland streambeds and huckleberry patches, have ambled off into the eastern woodlands.

Lauchlan Hamilton

Centuries of deadfall—dry leaves, conifer needles, bracken, crown after crown of sword fern, combined with the debris buildup of recent months, provide the fire with ample fuel as it picks up momentum across the broad plateau between the waters of False Creek and Burrard Inlet. Iron survey stakes and wooden posts marking future Vancouver streets and lots have been hammered into position over a wide area in the recent months by an eight-man survey crew headed by CPR land commissioner Lauchlan Hamilton. Carefully measured sections of future thoroughfare already named by Hamilton himself in honour of eminent individuals—heroes of battle, sea captains, politicians, men of business and science—lie directly within the fire's path. Several uncleared blocks of Drake Street, the closest in proximity to the roundhouse site, are the first to succumb.

Arthur Herring, the New Westminster wholesale and retail druggist frequently travels the twelve-mile New Road from New Westminster to Vancouver to deliver medicinal wares to the city. This being Sunday, he has brought along his family—with everyone looking forward to their customary afternoon picnic amidst the freshening seaside breezes of Burrard Inlet. As they descend the final hill towards Westminster Bridge, they are met with the usual vista of False Creek's tidal mudflats, log booms, fishermen's scow-houses, shacks and moored boats by the score. The view beyond is strangely obscured in a thick haze, much more concealing than the regular fog of the clearing fires.

Pulling up at Freeze's Stables to tether his team of horses, Herring asks if there is a fire about. "It's further to the west, in some brush clearing," he is told. Smoke is growing denser with each passing minute and great gusts of wind are whipping up. The horses are uncharacteristically restless. Feeling no small degree of apprehension himself, Herring decides that it would be a good idea to check on the well-being of his branch pharmacy on Water Street—but not before ensuring the safety of his family. Arthur and Frances Herring have lost three sons in recent years. They quickly retreat to the Bridge Hotel near south shore of False Creek, where Herring drops everyone off before heading back into the city.

At the intersection of Cambie and Cordova streets, lumberjack George Cary struggles with crewmen to keep another clearing fire under control. For the past few weeks, Cary has been working at this location, along the southwestern flank of Vancouver's central business district. While several stumps are still waiting to be dynamited, much progress has been made and the area is gradually taking on a more manicured look. Ed Cosgrove has recently completed construction on the three-storey, forty-room Regina Hotel, a block downhill at Cambie and Water streets. Access to water from the Regina's newly dug hand-pump well has proved to be a welcome convenience for fire control efforts. With this afternoon's fire at Cambie and Cordova proving to be particularly difficult to suppress, crewmen take rapid turns vigorously cranking the well pump and hauling full water buckets up the gentle slope.

■ 1:30 P.M.

The roundhouse clearing-site fire, having grown to monstrous proportions, is travelling at a high rate of speed across the isthmus. Having voraciously roared through the massive debris fields strewn atop the survey markers of Davie, Helmcken and Nelson streets, the flames continue their advance. Among the scores of fallen trees, the prevalent cedar and hemlock are flash fuels—resinous softwoods which ignite quickly and burn with ease. The giant specimens of Douglas fir—much prized by loggers and developers alike—are more resilient, surrendering gradually to the flames and burning hotter in the process. Tangles of underlying slash—long withered salmonberry, huckleberry, bracken, sword fern and salal, flare up like so many match sticks.

Sweating and cursing in the profound heat and smoke, the men at Cambie and Cordova diligently slam down shovels and aim full water buckets as spot fires continue to break out. A rogue breeze from the southwest has been steadily gathering strength, hampering their efforts at every turn. Strong and agile George Cary is called upon to run downtown and recruit some additional help. Rushing into one of the many Water Street saloons, he confronts Chief of Police John Stewart and delivers his message short and simple—"some men better come out and fight the fire!" Bemusement greets Cary in return. The Chief of Police, perhaps in an early state of inebriation, lounges back in his chair and waves his hand dismissively. "Oh yes, that will be all right!"[6]

At George Cary's persistent urgings, several men reluctantly push aside their glasses and follow him up to the clearing. What they see leaves them aghast. The wind has increased to gale proportions and is hurtling chunks of flaming branch overhead—some the thickness of a man's leg. Spot fires are breaking out in every uncleared patch of foliage. Men frantically pickaxe the hard-baked soil, in a last ditch attempt to create a firebreak along one of the semi-cleared sections of Cordova—but their actions are not having the desired effect. Flaming branches are hurtling past. Smoke is becoming so thick one can scarcely draw breath, the heat so intense that sweat-soaked shirts are beginning to steam. The deadly reality of the situation has dawned on everyone present—a massive fire, far greater than the one they have been attempting to control at Cambie and Cordova, is bearing down from the hillside. Blankets and shovels abandoned, the men run for their lives.

Ed Cosgrove desperately shouts and waves his arms for assistance at the Regina Hotel as several figures come scurrying down Cambie hill through a thick haze of smoke. Within moments, several volunteers join him in dashing from room to room, yanking new blankets off freshly made beds. A bucket brigade is formed to the well, ladders are positioned, and the Regina rooftop quickly takes on the appearance of a giant patchwork quilt, swathed from end to end with saturated bedding.

Lavinia Fisher has been told to "get out quick!"[7] There are shouts outside that a clearing fire is out of control and fast bearing down. With her husband Thomas away on business in Coal Harbour, Lavinia's first inclination is to dash about her Water Street home, gathering up armloads of clothing and blankets. The Fishers are proud to count themselves among the original pioneers of Burrard Inlet's south shore. Thomas Fisher helped the legendary John "Gassy Jack" Deighton build his first saloon in the area—much to the pleasure of thirsty Hastings Sawmill crewmen. On October 4, 1876, Fisher married his child-bride, fourteen-year-old Lavinia Cortez of Peru.

Trying her best, with much resourcefulness and making-do, young Lavinia has managed to carve a family life out of the wilderness for her brood of eight children. It had been no easy feat in the early years, when a single store at Hastings Mill was the only local base for supplies. Lavinia's eldest son, Thomas, is well into his teenage years, but the rest of the Fisher children—Walter, David, Isaac, Lawrence, Lavinia Rosella, Violet and baby Alfred, are all under the age of twelve and very much in need of supervision. Not wishing to raise any unfounded worries, Lavinia urges the older children to continue getting ready for Sunday school while she gathers belongings.

As per habit on sunny days, photographer Harry Devine has been scouting the city of Vancouver for prospective photo opportunities. His rucksack is loaded to the brim with the tools of his trade—a much prized E. and H.T. Anthony and Company camera with fixed aperture Darlot lens carefully wrapped in heavy, black focusing cloth, a supply of eight-by-ten inch wet photographic plates, a complete portable darkroom—plus a telescoping tripod over his shoulder. A native of Manchester, England, Harry Devine has recently arrived in the city from Brandon, Manitoba, with his business partner John Allen Brock. It has not taken him long to amass a sizeable collection of photographs—Vancouver street scenes, business establishments—even unique, panoramic views of the entire cityscape, captured from a boat out on Burrard Inlet. Devine is justifiably proud of his collection—so when news of the fire reaches his ears, he runs to his Cordova Street studio as quickly as possible, encumbered with over fifty-pounds of photography equipment.

Harry T. Devine

In a small office tucked in the back of W.E. McCartney and Bros. Chemists and Druggists, on the southwest corner of Water and Abbott streets, Alan McCartney busies himself catching up with paperwork. While his two brothers have taken up careers in dispensing pharmaceuticals, McCartney is a civil engineer and surveyor by trade, as well as the official accountant for Hastings Sawmill. The mill's complex finances require meticulous concentration. He is more than grateful to have use of the quiet room in his brother's establishment, well away from the ever-present roar of the saw blades—so when he first notices an abnormally thick amount of smoke billowing past his window, he is more annoyed than perplexed. If there is a clearing fire in the area, it is much too close for his liking.

A brief look up Abbott Street sends Alan McCartney flying back inside to the drug store shelves. The McCartney brothers operate their business as a branch of Arthur Herring's wholesale and retail drug establishment in New Westminster. Since opening in January, they have accumulated a sizeable stock—drugs, chemicals, patent medicines, toiletries, perfumery—all manner of tonics and elixirs, much in demand by Vancouverites not only trying to cope with a variety of ills, but seeking any small indulgence of personal hygiene in a frontier settlement. The entire supply could well be at risk if any of the sparks and firebrands swirling past should land upon the roof. McCartney quickly begins packing bottles into boxes and carrying them across Water Street to the Burrard Inlet shore. It is not long before he is joined by Arthur Herring, anxious to check on the wellbeing of his Vancouver store.

The Boultbee family lives in a comfortable shoreline house on the eastern outskirts of Vancouver, where False Creek peters into tidal mudflats. Magistrate John Boultbee shares his downtown business office with Charles Gardner Johnson, owner of a growing business in shipping and insurance. In addition to being neighbours and best of friends, the Boultbee and Gardner Johnson families are in-laws, Minnie Gardner Johnson being John Boultbee's sister.

In the heat of the Sunday afternoon, both families are enjoying refreshments at the Boultbee household. A delighted Helen Boultbee, one day short of her ninth birthday, has been given the responsibility of walking into town to purchase eggs for a simple evening meal. To everyone's surprise, she returns earlier than expected, without having made a purchase.

"Some men told me that there's a bad fire! People are running away from it!"[8] she announces excitedly.

Everyone rushes outside at her words. A thick black cloud of smoke is billowing skyward over the city. Charles and John set off on a run up Westminster Road. Elizabeth Boultbee and Minnie Gardner Johnson spring into action, racing about their dwellings and trundling loads of household goods and valuables down to the water's edge of False Creek. An incoming tide has just begun to lap at the edges of the Boultbee family's raft, moored near the shoreline.

■ 1:45 P.M.

George Cary runs from the Cambie and Cordova clearing site towards Burrard Inlet, more concerned with staying alive than joining efforts to save the Regina Hotel. Smoke and cinders are hurtling past him and flames are licking at the walls of a nearby shack. He hears a frantic call. A bewildered Mrs. Irwin stands transfixed, two children clinging to her skirt, while her husband John beats at flames outside of their clapboard home.

"Look after my wife and kids!" he yells to Cary.

Mrs. Irwin insists upon putting on her hat. Not quite certain of what to do, Cary shepherds the family down to the foot of Cambie Street, where the CPR has been filling in a section of waterfront with gravel and pilings. Nearby, a group of men on a float are quickly poling away from shore. They are still within earshot and Cary yells, "I've got a woman and children!" Much to his dismay, the men have no intention of waiting. Seeing another raft being maneuvered out less skillfully, Cary clambers out over the fill and plunges waist-deep into the water. This time determined not to be ignored by the occupants, he slogs forward and seizes a corner of the vessel. When one of the men aboard angrily swings at him with a piece of two-by-four, Cary quickly decides that enough is enough. He heaves himself aboard, wordlessly grabs the two-by-four and hurls it out of harm's way. With the help of an incoming tide, he backs the raft into position for Mrs. Irwin and her children to clamber aboard. There is little space and the craft is bucking wildly in the waves. The children are crying in terror, but above the din, George Cary hears a distraught Mrs. Irwin make a simple statement: "I'd rather see them drown than burn!"[9]

At the west end of Water Street, not far from unbroken forest, stands the two-storey home of new Vancouver residents, Emily and Alexander Strathie. Emily takes great pride in the dwelling, which her husband Alexander has just completed. During the past few weeks, the Strathie parlour has become a refuge of sorts, for fashionable ladies new to Vancouver. Here, Emily invites her guests to sink into comfortable brocade chairs and enjoy a soothing cup of tea and a piece of cake or toast, after their arduous journey over the bumpy wagon road from New Westminster. It is a goodwill gesture to which she is dedicated, leaving Sunday the only spare time for self-indulgences like an afternoon stroll to the nearby military reserve, soon to be known as Stanley Park.

Today, a friend by the name of Mr. Haslam is joining the Strathies. While he and Alexander venture outside to make inquiries about the clearing fires, Emily busies herself upstairs. High fashion being out of place in a city that is more bush than thoroughfare, she changes from her Sunday finery to a cheap print dress and pair of slippers. A casual glance toward her bedroom window makes Emily stop and stare, open-mouthed. A giant tongue of flame has just swept along the narrow alleyway between her house and the one next door.

Was she hallucinating? She hears Alexander, yelling from the stairwell. "Fire, Emily! Come quick! Don't waste any time!"[10]

Dazedly, Emily finds herself grasping the closest thing within her reach—her husband's hat. In the next instant, the bedroom window crashes in, shards of glass spraying over the floorboards. Shaken to her senses, Emily runs downstairs.

"Run for the waterfront!" Alexander yells, already in the process of hauling a trunk across the parlour floor and out onto the veranda. The Strathie trunk is full of fine clothing, jewelry and other treasures, "proper enough for a city but quite unsuitable for wear in the rough and ready old Granville,"[11] as Emily liked to put it. With Emily not far ahead, Alexander drags the trunk down to the beach and heaves it atop one of the many barnacle-encrusted boulders embedded in the Burrard Inlet shore.

Alan McCartney and Arthur Herring, exhausted from carrying heavy caseloads of drug store stock to the beach of Burrard Inlet, go their separate ways—Herring more than anxious to return to his family. At the nearby Hole in the Wall Saloon, a crowd of lunch hour patrons look up from their drinks indifferently as McCartney barges through the door and yells at them to run for their lives. The room erupts with laughter. In frustration, he leaves, almost running squarely into Henry Abbott, the General Superintendent of the Canadian Pacific Railway's western division. Abbott has been walking up his namesake street, trying to determine the cause for the unnatural amount of smoke.

Henry Braithwaite Abbott

"Things look blue!"[12] he remarks grimly. Drawing a pencil and piece of notepaper from his pocket, he scrawls out a message and hands it to McCartney.

"Good for all the pails you can find. H. Abbott, Gen Supt. CPR."[13] Coming from Henry Abbott, the note is as good as any amount of cash. Clutching it in his hand, Alan McCartney races on to F.X. Martin's General Store, where he knows that there is a good stock of hardware. People are now filtering out of the saloons, peering curiously about. A general alarm is finally beginning to circulate from one building to the next.

Dr. William McGuigan sips a refreshing drink in the bar of the Deighton House Hotel, passing the afternoon in idle conversation with his friend, Barrister John Blake. Deighton House, in the very hub of the city at the corner of Water and Carrall streets, was built over a dozen years ago by Vancouver's unofficial founder, John "Gassy Jack" Deighton—who provided much entertainment for patrons with his colourful stories and ever-present gift of the gab. Gassy, though long gone, has left his indelible stamp with the city's nickname of "Gastown" still in wide use. Also retaining the name of its founder, Deighton House continues to accommodate the city's elite with its comfortable parlours and extensive dining rooms.

Conversation in the bar is instantly silenced when Andrew Muir bursts through the door, a look of utter dismay on his face.

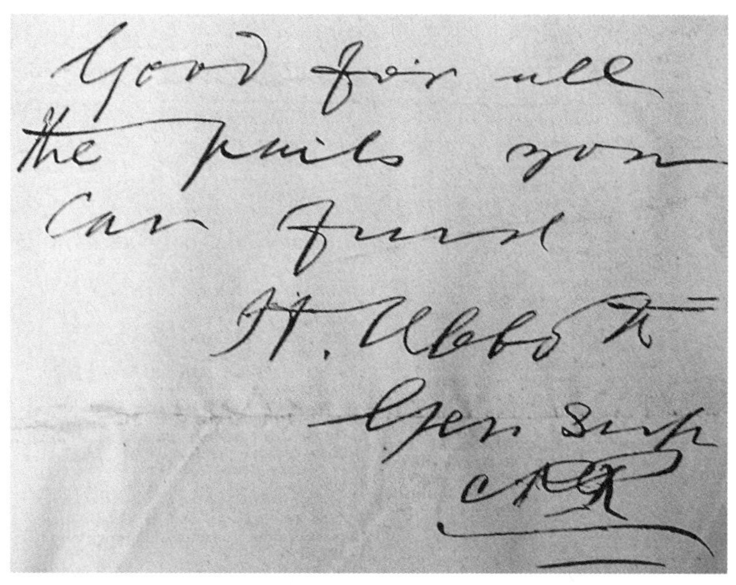

Good for all
the pails you
can find
H. Abbott
Gen. Supt
CPR

"Good for all the pails you can find."

"Birdie Stewart's house is on fire!" he cries.

"Birdie Stewart's house?"

Various bar patrons exchange carefully orchestrated glances, betraying nothing. Birdie Stewart's house, one of Vancouver's more established brothels, sits near the Methodist Church at the western end of Water Street.

"Yes! The entire town is going to burn!"[14]

Peering outside, Dr. McGuigan frowns. Smoke is thick in the air, to be sure, perhaps thicker than usual, but there are no flames in evidence. Bidding good afternoon to his companions, he returns to his room in the Sunnyside Hotel, directly across Water Street from Deighton House. With no immediate sense of sense of urgency, he pulls up a comfortable chair and settles in for an afternoon snooze, idly daydreaming of a long lost love back in Ontario.

William Gallagher, tired and breathing heavily after his hectic morning at the roundhouse clearing site, reaches his office at the north end of Columbia Street. A thick veil of smoke has followed him all the way back to town. Gathering up several "Percival and Gallagher" accounting books and stuffing thick wads of cash into his pockets, he ponders his next move. Should he saddle up a horse and return immediately to the clearing site to confirm that the crew are regaining control, or await further word? Glancing out of a south-facing window, Gallagher notices several people running in various directions across Maple Tree Square, apparently in a state of agitation. The boughs of the big old maple tree—a popular meeting place at the busy juncture of Carrall, Water, Alexander and Powell streets, are swaying fiercely in the wind. Gallagher decides that it would be prudent to walk up to Deighton House and make inquiries.

Dr. William McGuigan

Hugh Campbell, an athletic twenty-one-year-old, is one of forty young men who attended the inaugural meeting of the Vancouver Volunteer Hose Company Number 1 on the evening of May 28 at George Schetky's men's clothing store. Storekeeper Sam Pedgrift was elected fire chief, and monthly dues were established at "2 bits,"[15] or twenty-five cents. With little to show for in the way of contemporary firefighting equipment, the company is focusing its training sessions on preventative strategies. Buckets and shovels are kept in good supply. The locations of Vancouver's water sources—creeks, wells and the two flanking inlets—along with their best access points, have been carefully noted. With quick response, a small fire can be easy enough to extinguish. Everyone is in agreement that if a major fire develops in one of the city's larger structures, at present, little can be done. Good care must be taken to stifle small flames before they become unmanageable.

Whiling away another hot afternoon at the Sunnyside Hotel, Campbell hears the shouts of "fire" reverberating from Water Street. He ventures outside and quickly receives confirmation of the news that every member of Hose Company Number 1 has feared most. It's a big one—out of control, and advancing fast. Campbell knows instantly what he has to do. Making a direct run for Tattersall Stables and Livery, he joins several others attempting to harness and blindfold the horses. For all its strength and usefulness, a horse sensing fire will often make straight for the perceived safety of its stall. After hitching up one of the nervous beasts to a two-wheeled butcher cart, Campbell quickly leads him up Water Street to Scoullar's General Store. Located at the western edge of town between Abbott and Cambie streets, Scoullar's is one of Vancouver's well-stocked hardware stores. Among other items on its shelves are several wooden crates of stumping powder—a low grade dynamite in wide use for blowing up stumps at the clearing sites. The men of Hose Company Number 1 are under strict orders to remove explosive materials from the city as quickly as possible in the event of a fire.

Scoullar's is a frenzy of activity as buckets and shovels are yanked off shelves and unopened crates of newly shipped hardware are trundled out to Water Street. Hugh Campbell loads as many crates of stumping powder as he can safely fit onto the butcher cart, and momentarily ponders his dilemma. To the west, Water Street becomes a narrow, meandering trail through largely uncleared forest. To the east, the city appears to have gone mad. People are pouring out of their homes and businesses. Shouts and screams fill the air as an atmosphere of panic takes hold. The idea of wheeling a cargo of explosives amongst a frenzied crowd seems ludicrous, but it is clear that he has no choice. Shouting for space, Campbell begins a perilous rush east along Water Street, clutching the reins of his blindfolded horse, butcher cart jostling wildly behind. He decides to head for the only logical destination he can think of—Hastings Mill, at the opposite end of town.

With fire bearing down fast, Lavinia Fisher is forced to abandon her efforts to save household goods. She scoops up baby Alfred and toddler Violet in her arms and yells to the older children to run for Hastings Mill. In the rapidly changing Vancouver street scene, it is a landmark which she knows they will be familiar with. Nine-year-old Walter is put in charge of four-year-old Lavinia Rosella. Wind blasts at their backs as the Fisher family make their way up Water Street. People are running crazily in every direction. Horses pulling wagonloads of goods thunder past, stirring up clouds of dust in the smoke. Amidst the throng, Lavinia desperately tries to keep track of her children. They are keeping up as best they can, but in all the commotion, Walter and Lavinia Rosella have become separated from the rest. Backtracking is not an option. Clutching baby Alfred and Violet, calling her missing children all the while, Lavinia continues on to Hastings Mill with the rest of her family.

When James Ross, co-editor of the *Vancouver Daily News*, learns from his business partner N. Harkness that his home, wife and baby daughter are in imminent danger, he races to their aid. Ross, originally from Belleville, Ontario, has made his way west to Vancouver after a period of employment with the *Winnipeg Free Press*. Vancouver's growth has been so rapid through the early months of 1886, that no less than three newspaper companies have been established to chronicle the pace: the *Vancouver Weekly Herald and North Pacific News*, the *Vancouver Daily Advertiser* and the *Vancouver Daily News*. The latter is the most recent, its inaugural issue appearing on the morning of June 1, with an introductory editorial:

> Our aim will be to improve as the town improves, to keep pace with its progress, however rapid, to make the paper at all times worth its price.[16]

The *Vancouver Daily News* printing plant—a small building on the west side of Abbott Street between Water and Cordova—is rapidly being enveloped with smoke. A half-completed issue of Monday, June 14th, lies abandoned on the press.

Laundryman Wah Chong's lines of freshly washed bed sheets and table linens flap wildly in the driving wind. When Canton became the sole port permitted by the Chinese government to engage in foreign trade and commerce, thousands of local peasants from the eight rural counties of Guangdong Province made their way to the New World, Wah Chong and his family among them. While most Chinese residents of "Hahm-sui-fau"—"Saltwater City"[17] (as they call Vancouver) are employed at Hastings Sawmill, Wah Chong has defied convention. Having arrived from San Francisco with his wife, three sons and two daughters in 1884, he has established a well-patronized washing and

Wah Chong family outside their Water Street laundry business, 1884.

ironing premises in the very heart of Vancouver's Water Street business district. Fine granules of soot collect on the white sheets, but no one is pausing to empty the lines.

It is rumoured that One-Armed John Clough (aptly named for the missing right arm long lost in a dynamiting accident), had been imprisoned for drunkenness so many times that the Police Commission decided " it would be cheaper to take him on staff."[18] As the city jail keeper, it is Clough's responsibility to monitor the hapless individuals he can relate to so well—those who have been incarcerated, usually to sober up after saloon behaviour gone too far. Vancouver's jailhouse, a little wooden shack with four cells off Water and Carrall streets, is nothing compared with the maximum security B.C. Penitentiary, built in New Westminster in 1878, but it serves its purpose well for short-stay occupants.

John Clough

Clough warily eyes the ever-thickening smoke drifting past. Only one cell of the jailhouse is occupied at present. If it was not his responsibility to monitor that solitary individual, he would be sorely tempted to head for the beach. Something is indeed wrong. Minutes later, Police Chief Stewart runs up and confirms his suspicions, authorizing the jail keeper to release the prisoner and run for his life. John Clough needs no second invitation. He unlocks the cell door, then, before taking to his heels, he grabs up a hammer, correctly speculating that it will be a useful tool in days to come.

When word reaches city clerk Thomas McGuigan that an "out of control" clearing fire is bearing down on Vancouver, he immediately knows what he must do. He is officially entrusted to the care of the Vancouver City Archives. Of utmost importance are the minutes, including those from the very first Vancouver City Council meeting, held on May 10—a delicately handwritten, irreplaceable record of that great milestone in Vancouver's short history to date. Additionally, there are new city bylaws, police records, letters, the first city voter's list and results (however falsified they were rumoured to be), of the first mayoral election. Everything is under the direct responsibility of the City Clerk.

Thomas McGuigan races to the water's edge, his valise stuffed full of the precious documents. A scow floats in the shallows, its owner making ready to cast off. McGuigan eyes the man suspiciously, wondering if he has found someone he can trust. There is no time for indecision, and he is anxious to try to rescue some of his own personal effects. Abandoning the valise to the care of a total stranger, he runs back into town.

Thomas F. McGuigan

The Miller household, two lots west of Abbott Street between Cordova and Water, is a crowded one. Nine Miller children share two bedrooms, while a living room, kitchen and ever-present rocking horse on the front veranda comprise the rest of the abode. Longtime resident of Burrard Inlet's south shore, Jonathan Miller formerly served as the first police constable of Granville township. Now in his fifties, he has taken up a more sedentary position as Vancouver's postmaster. Miller's second-eldest daughter, Carrie, is about to get changed for Sunday school when urgently told by her father that she and the rest of the family must run away quickly. She spontaneously grabs the first item of clothing her hands come across—a thick winter dress. Sweltering profusely in the heat and smoke, Carrie takes charge, leading her distraught mother and siblings up Water Street. Jonathan Miller and his oldest sons remain behind to carry as much clothing, furniture and bedding as they can manage to a nearby vacant lot, in hopes that it will somehow escape the flames.

Sunday, June 13, 1886

On a long float jutting into Burrard Inlet at the north end of Carrall Street, a throng of people converge upon Andy Linton's boathouse. New Brunswick-born Linton, unofficially recognized as Vancouver's resident boat builder, has painstakingly constructed a small fleet of rental vessels from the finest available local materials. Today, several of Linton's boats have been rented by individuals attending a native gathering at the mission on the north shore of Burrard Inlet. All he can do is shout brief instructions above the din, as the rest of his fleet is rapidly and recklessly commandeered. A handsome new eighteen-footer, his pride and joy, leaves the float with a not particularly competent-looking crew at the oars. No one has a definite destination in mind—just to get out on the water quickly, and well away from shore. Some people have boarded the boats with blankets tucked under their arms. It is impossible to hoist sail against the driving wind, but the blankets resemble ragtag versions of sails as they are soaked with seawater and held aloft to shield against the volley of sparks flying overhead.

Jonathan Miller

Some Vancouver citizens are caught in awkward situations as the fire bears down upon the city. Mrs. Pedgrift settles blissfully into her warm tub for a Sunday afternoon bath. The children have been bundled off to Sunday school, and an hour's serenity is at hand. Pedgrift's General Store on Powell Street specializes in shoes but carries all manner of dry goods. Sam Pedgrift and his wife, ever mindful of appearing clean and respectable for their customers, have recently added the backyard bathhouse to their premises.

After learning of the fire, George Allen, the Pedgrift's young apprentice shopkeeper, knows exactly where to find his manageress. He races to the bathhouse and pounds hard on the door.

Andy Linton

"Mrs. Pedgrift, you must get out quickly!"

"I'm in my bath!" comes the indignant reply from within.

"It doesn't matter what you're in, you must get out quickly, or you'll be burned up!"[19]

A glaring Mrs. Pedgrift finally emerges. Any chastisements for her employee quickly die on her tongue as she realizes the seriousness of the situation. Moments later, she is racing up Powell Street in search of her children, not in her ideally chosen state of dress.

When Harry Devine arrives at his studio, his parents and elder sister Annie are already hard at work, loading fragile negative plates into a wheelbarrow with as much care as they can afford. With wheelbarrow piled high, the Devine family joins the crowd running down Carrall Street in the direction of False Creek. John Devine deftly maneuvers the wheelbarrow in and out of human traffic, while Harry, loaded down with his backpack, struggles to keep up. The Devines quickly become separated amidst the noise and confusion of the crowd.

Sunday, June 13, 1886

False Creek, primarily a tidal inlet, takes a sharp dog-leg to the north at Carrall Street—a fortuitous escape route for those fleeing the fire from the centre of town. Along the shore, vessels of every size and description are being deployed—everything from the tugboats of the Royal City Planing Mills to dugout canoes from Snauq. Harry Devine manages to leap aboard a small raft, while further down shore, the rest of his family climb aboard a barge, John Devine carefully balancing the wheel-barrow-load of negative plates.

In one arm, Jack Alcock cradles a small valise as he makes his way to False Creek. The other arm is wrapped protectively about the shoulders of his Uncle Small. The elderly man is struggling to catch his breath and maintain his footing. A crack shot huntsman, Alcock has stuffed his valise with cartridges, in hopes that he might be able to supply food for the family in days to come. The cartridges are heavy and, with smoke thick in the air, both men cough uncontrollably as they stagger along. Finally, with little regret, Alcock lets the valise slip from his grasp. It lies unguarded and alone amidst a wilderness of stumps and half-cleared ground.

A few individuals remain blissfully ignorant of the danger surrounding them. Not yet two months old, Margaret Florence McNeil, first white baby of the incorporated city of Vancouver, slumbers peacefully as her parents run with her to False Creek. Two-week-old Frederick Macey lies in his cradle, unperturbed, while the rest of the Macey household scramble to collect family belongings. As grown adults in ensuing years, Margaret and Frederick will relate their fire experiences from vivid descriptions provided to them, rather than memories of their own.

Christina Reid

Christina Reid, eight months pregnant with her third child, helps her seven-year-old daughter Jemima dress for Sunday school, when a fierce pounding echoes through the house from below. Someone is banging urgently and repeatedly on the front door of the Reid family's Dupont Street home. Duncan Reid, roused from his afternoon nap, groggily makes his way downstairs. He opens the door to a stranger, who is beside himself with agitation as he yells, "Get out quick or you'll be burned in a few minutes!"[20]

Fire alarms have happened before on Dupont Street, with its close proximity to the clearing piles. Whenever they occur, the Reid family rule is for Christina and the children to head for False Creek, while Duncan attends to the livestock. The fastidious Reids have practised fire drills many times over—actions being carried out in precise, orderly fashion, with everyone reuniting at home once the danger was past. Nothing, in all of their weeks of preparation, has quite readied the family for what is happening now. Tongues of flame are licking at dry grass in neighbouring yards. Duncan Reid realizes that even if his very expectant wife were able to run, she would never be able to manage young Jemima and baby Minnie.

Father Henry Fiennes-Clinton, having finished a quick bite of lunch, is preparing to welcome children back for afternoon Sunday school at St. James Anglican Church. Preaching the gospel in a frontier community like Vancouver has taken some getting used to. The young Englishman is fresh from a five-year position as curate at St. Mary Magdalene's, in the Yorkshire town of Bradford. Lately, however, he is beginning to enjoy the rewards of his dedication, as he describes on April 3, 1886, in a letter to his sister-in-law Lucilla. "We have got a nice little church here and don't we have it full now on a Sunday."[21] The use of a horse, and a small sailboat for pastoral duties across Burrard Inlet in the north shore town of Moodyville, have also been provided through the church. St. James Church occupies a beautiful position on the Burrard Inlet foreshore, a short distance west, but within the property boundaries of Hastings Mill. Its little iron bell, while by no means comparable to the resounding giants of Yorkshire's grandest cathedrals, serves its purpose well in calling parishioners to worship every Sunday.

Father Henry Fiennes-Clinton

Between pulls of the bell rope, Father Clinton hears loud voices from Alexander Street. Coming outside to investigate, he encounters a small, panic-stricken group running past the church, in the direction of Hastings Mill. One glance to the west makes no further explanation necessary. Racing back into the church, Father Clinton grabs the bell rope once again and proceeds to haul on it relentlessly—a call of alarm, replacing the call to worship.

■ 2 P.M.

Not far from where Emily and Alexander Strathie have deposited their family trunk, two lumberjacks are hard at work in knee-deep water improvising a raft from cut lumber. The boards have been carefully stockpiled in recent days for use in building a tailor's shop. Without any ready access to hammers, nails or ropes, all the men can do is lay the boards criss-cross, in hopes that somehow the makeshift craft will hold together. Emily and Alexander join in the effort, clamping their hands over one wildly bucking section after another.

Others soon converge on the raft as more and more people flee the driving flames. Louisa Wilson staggers down, her arms tightly clutched about the shoulders of her father, James Morrison. Morrison is suffering from gout, and can barely stand, let alone walk. Several helping hands materialize to carry him through the shallows and hoist him aboard. Before long, the criss-crossed boards have a passenger load of fifteen men and two women, struggling desperately to stay afloat. Reverend Joseph Hall, along with fourteen-year-old Sunday school teacher Jessie Greer, clamber on. Reverend Hall clutches a length of lumber with one hand and holds his Sunday Bible aloft with the other,

St. James Church, 1886.

as the lumberjacks gingerly attempt to pole the craft to deeper waters. In no time, most of the passengers are waist-deep as the boards yawn apart. Louisa Wilson cries for help as her father submerges to his neck.

Anchored further offshore are two dinghies—the kind often towed behind the pleasure boats that frequently ply Burrard Inlet. Alexander Strathie suddenly announces that he is going to try to swim to one of them, thereby lightening the load for the others. He urges his wife to come with him, perhaps inwardly reasoning that he can somehow tow her along, as Emily is a non-swimmer.

"You cannot make it!" she argues.

Alexander persists, but Emily is resolute. "You go alone! You will never be able to make it with me, we shall both drown!"[22]

This matter-of-fact statement evidently convinces Alexander

to remain with his wife. As flames and smoke soar over their heads, the Strathies and their companions cling to individual pieces of lumber as the raft washes apart.

"What in the world are you doing!" a voice bellows from the doorway.

Dr. McGuigan opens his eyes to find John Blake staring at him incredulously.

"What's the matter, Blake?"

"Why, the town is burning down all around us! The roof next door is on fire! Get your stuff packed as quickly as you can and get out or you'll be burned right here with your boots on!"[23]

Snapped out of his reverie, Dr. McGuigan throws some clothes and medical books into a trunk and begins to haul the load down the Sunnyside Hotel staircase. In that same moment, flames burst through the ceiling above his head. Abandoning the trunk, he races outside, to confront a stampede of townsfolk hurtling their way along Water Street. Bent low with a handkerchief clasped over his mouth, Dr. McGuigan dodges his way through the crowd towards his Carrall Street surgery. A fierce wind tears down the street, "50 miles per hour at least,"[24] the doctor surmises to himself in awe. With great effort, he manages to reach the surgery, grabs up a few of his most frequently used operating instruments and, on the spur of the moment, a small bag of opium. Dashing back out to the street, he again meets up with John Blake, burdened down with a load of legal papers. With Vancouver's main thoroughfares rapidly becoming an inferno, the two men run eastward on Powell Street, breathing laboriously under the combination of their heavy loads and smoke-filled air.

Manager Harry Hemlow gazes in dismay at the lobby of his beloved Sunnyside Hotel. It seems unthinkable that the grand structure, pride of Vancouver and home to the city's most elite tenants, is about to be destroyed. The ridiculous irony is that the hotel literally stands over water, atop wooden pilings machine-driven deep into the Burrard Inlet bedrock. At high tide, Burrard Inlet seawater washes beneath the Sunnyside's large new addition. Only short days ago, newly elected Mayor Malcolm MacLean had delivered his victory speech from the Sunnyside balcony to a crowd of cheering supporters below. Water Street is once again jammed with people, but now the mood is far from jubilant.

The Sunnyside's popular shoeblack and porter, Joe Fortes, shepherds Sunnyside tenants Jessie Ross and her eight-year-old

Granville, 1885. The large white building left-of-centre is the Sunnyside Hotel.

son Donald down the hotel stairs and out to the water's edge, where all manner of rescue craft are converging. Big and powerful Fortes, a native of Trinidad, has been working at the Sunnyside for the past several months. He has quickly established a reputation for his jovial good nature, coupled with the strength and prowess of a star athlete. After seeing to the safety of Sunnyside guests, Joe Fortes and Harry Hemlow climb aboard a small scow and paddle away from shore as fast as they are able, using only their hands for oars.

Captain Thomas Jackman, assistant to Harry Hemlow, is intent upon rescuing his trunk from the burning Sunnyside. While heaving it aboard a nearby scow, he is confronted by Mr. Terhune, personal secretary to Henry Abbott. Much to his surprise, Mr. Terhune produces three large bundles of cash notes from beneath his waistcoat. He has sprinted from the Canadian Pacific Railway offices on Carrall Street to ask if the money can be placed in the Sunnyside safe. The two men return to the hotel and together jam the CPR cash into the almost full safe, swing the door shut, and run for their lives.

William Gallagher stands aghast, watching flames soar from the rooftop of the Sunnyside Hotel. Clusters of terrified men, women and children are fleeing up the middle of Alexander Street, having abandoned the wooden plank sidewalk adjacent to the thoroughfare. Fire is racing along the bone-dry planks faster than a grown man is able to run. In desperation, many people have chosen to dart between the burning buildings and throw themselves headlong into the waters of Burrard Inlet. Gallagher quickly chooses this option for himself, plunging into the inlet just beyond the office of Percival and Gallagher. Smoke billows past, so thick that he can barely see his hand in front of his face.

He soon discovers that by immersing himself to his shoulders, he can inhale cool fresh air close to the surface of the water. Jockeying for balance amidst the submerged boulders, Gallagher can see only occasional glimpses of his burning office between clouds of smoke.

As they make their escape, many Vancouverites are laden down with all manner of items, ranging from the practical to the absurd—overstuffed satchels, winter coats, blankets, cooking paraphernalia—even a bird flapping wildly in its cage. William Kent stashes his trunk and valise in a newly-made CPR culvert, then runs back to his El Cisne Hotel for five hundred cigars and some billiard balls. Hiram Scurry trudges along, delicately attempting to balance a looking glass atop his barbershop chair. Saloonkeepers roll kegs of whiskey into the waters of Burrard Inlet, where they pitch and bobble drunkenly amidst a rapidly expanding flotilla of rescue craft. A well-meaning Arthur Herring, having stopped to help a man to gather up pots, pans and dishes in a cloth, asks "Is that it?

"Yes," the man replies.

"Then drop it and run for your life!"[25]

The roof of Graveley and Spinks real estate office, at the intersection of Carrall and Cordova, has caught fire. Young realtor Walter Graveley dashes about inside, grabbing up his company cashbook and ledger—valuable records of a booming business in Vancouver land sales. Vancouver has been a dream-come-true for Ontario-born Graveley. On March 6, 1886, he was among the purchasers of the very first city lots being sold by the CPR. Over the ensuing months he had tramped through miles of forest, envisioning future neighborhoods of cleared, graded streets with stately homes and manicured gardens. Buyers were lining up and prospects had never looked brighter.

Smoke sears Graveley's lungs. Cries of panic reverberate up and down Carrall Street. As he steps outside with cashbook and ledger tucked under his arm, Graveley finds himself performing an absurd action—he turns the key and locks the door of his burning office. Casting a despairing glance towards the centre of town, he can see the Sunnyside Hotel, home to all of his other worldly possessions, going up in flames. There is nothing to do but make a run for safety.

"Give me a hand with these children!"[26]

Robert Balfour, proprietor of the Burrard Hotel, is desperately trying to shepherd a small group of frightened youngsters up Carrall Street. Walter Graveley quickly hoists a tearful young lad atop his shoulders, and the group makes a run for False Creek.

The two-storey Ferguson Block stands at the busy corner of Carrall and Powell streets—Vancouver's first building to house multiple businesses. In his Canadian Pacific Railway office on the second floor, land commissioner Lauchlan Hamilton frantically scrambles to gather up the survey maps he has painstakingly created over the past months. In all his years of survey work—laying out the townsites of Regina, Swift Current, Moose Jaw and Calgary—Hamilton has met with many challenges, but nothing on the scale of a massive wildfire. Clutching a sheaf of maps, with his survey level tucked under his arm, he begins a mad run to False Creek.

By the time Charles Gardner Johnson and John Boultbee have made their way to Carrall Street through a crowd of people rushing in the opposite direction, structures to the right and left are in flames. Stumbling inside their office, they peer around the smoke-filled room. Paint is blistering on the opposite wall and bright flames are dancing past the north-facing window. A small tin box atop one of the desks catches Gardner Johnson's eye. It contains his sailor's papers. That small box has accompanied him all over the world, on a host of adventures and misadventures. He hastily stuffs it in his pocket. Across the room, John Boultbee is wringing his hands over the endless volumes of court proceedings and other legal documents that have accumulated over the course of his career. It would take hours to sort through and extract the most important papers.

Charles Gardner Johnson

Both men quickly realize that any attempts to linger and gather up more belongings would be futile—and likely fatal. Together, they run out of the office and turn a corner onto nearby Hastings Street. Hastings, like nearby Cordova, is still only partially cleared of forest undergrowth and stumps. While pondering their next move, the men hear a desperate cry for help. George Bailey, bartender of the recently-opened Balfour Hotel, is trying to extinguish several small fires licking at the rear of the building. Despite any misgivings, Gardner Johnson and Boultbee join him in swatting at the flames with wet blankets. Samuel Macey soon joins them—a stranger to the three but ready to offer his assistance. The men go inside, where bottles of brandy, whiskey and other spirits line the bar shelves, their contents glowing brightly in the face of orange flames.

Thomas McGuigan races down Columbia Street, clutching a load of personal belongings. He quickly realizes that he will make far better headway without being so encumbered. At the south-east corner of Powell and Columbia, Vancouver grocers David and Isaac Oppenheimer have recently begun construction of a large new warehouse. While the framework of the building is yet to be erected, a deep cement root cellar is in place. The cellar is vast—nearly a full city block in length and McGuigan suspects that being made of cement, it will prove the ideal place to store his belongings. He tosses the load into the cellar and continues to run for False Creek.

Jackson Abray owns a boarding house and restaurant at the northwest corner of Powell and Columbia streets. It is typically a busy establishment and he has forty-two men from the clearing crews seated and enjoying lunch. Above the conversation in the crowded room, Abray hears shouts from outside and ventures out his front door to investigate. The Miller family is running past, clearly distraught. One of the daughters yells to him that the city is on fire. Appalled, Abray hurries back inside and urges his guests to "move at once!"[27] With the help of his cook, Mr. Anderson, he races about, casting aside plates with half-finished meals and gathering up the silverware, table linens and bed sheets from the adjoining rooms.

Jackson Abray

A small raft is moored at the Burrard Inlet shoreline a block away, which the men quickly load with rescued goods. Above the roaring wind, calls for help reach their ears from the direction of the *Vancouver Weekly Herald and North Pacific News* printing plant, not far from the restaurant. William Brown is desperately struggling to save what he can of the apparatus used to publish Vancouver's first-ever newspaper. Jackson Abray grabs up a box containing a load of lead type—every letter of the alphabet in a variety of sizes and fonts. William Brown clearly has no desire to see the collection dumped into Burrard Inlet, where the type would be hopelessly lost to the waves and corrosive seawater. Mr. Anderson has a heart condition and is breathing heavily, but insists on hurrying back to his room at the boarding house to retrieve his Masonic jewels. With their arms full, the men emerge onto Powell Street, not at all certain of which direction to head next.

A run to the stable yard proves futile for Duncan Reid—the speed of the fire makes any thought of hitching up a wagon sheer folly. A drainage ditch runs alongside Dupont Street at the edge of the Reid property. Its water level, normally knee-deep or higher after a few days of rain, has evaporated to a few stagnant puddles, green with algae and alive with black flies and mosquitoes. With no other immediate means of escape, Duncan urges his wife and daughters to lie down in the ditch and splash themselves as thoroughly as they can. Christina Reid, clutching her daughter's hands, a parasol and six silver teaspoons hastily grabbed from the parlour, determinedly wades into the deepest puddle she can find. Jemima starts to cry as thick mud oozes over the lacework of her Sunday dress. Duncan disappears momentarily, and returns with an armful of stable blankets. After soaking them as thoroughly as he can in ditchwater, he throws them over his prone family. With his clothing beginning to smolder, Duncan races back and forth along the embankment, wetting and re-wetting the blankets. Steam rises from the coverings over little Minnie, and she whimpers as a glowing ember burns through and singes her baby hair. Christina gasps as a resounding boom shakes the ground all around them. An intense buildup of heat and pressure has blown the roof off the nearby soda water factory. Jemima covers her ears and sobs in terror.

Hugh Campbell races up Alexander Street with his cartload of dynamite. Sparks and flaming debris sail over his head, propelled by a wind quite unlike anything he has ever known before. His objective, the Hastings Mill townsite, is swarming with what could best be described as orderly panic. The mill whistle, commonly used to signal shift changes, is being sounded again and again in alarm. Boats are rapidly being deployed from the dock area, while men bearing armloads of blankets and buckets race in every direction. Only one ship is moored at the mill at present—the *Southern Cross*, under the command of Captain Cox.

Over the years, Captain Stamp's original mill site has swollen to include a store, cookhouse, bunkhouses, a school, library, doctor's office, blacksmith shop, sawdust sports field and several residences—virtually a city within a city. The giant refuse burner built alongside the sawmill roars with activity day and night. A small shed opposite the cookhouse houses the community's one and only fire engine. Built in 1850 by John Rogers of Baltimore, Maryland, the third-class manual pump nicknamed "Telegraph," served Victoria from 1858 to 1862. In 1873, the re-built machine was sold to Hastings Mill—where it had become something of a local laughing stock. "We Git Dar!"[28] was the motto of the Dark Town Fire Brigade, a group of mill workers in blackened faces

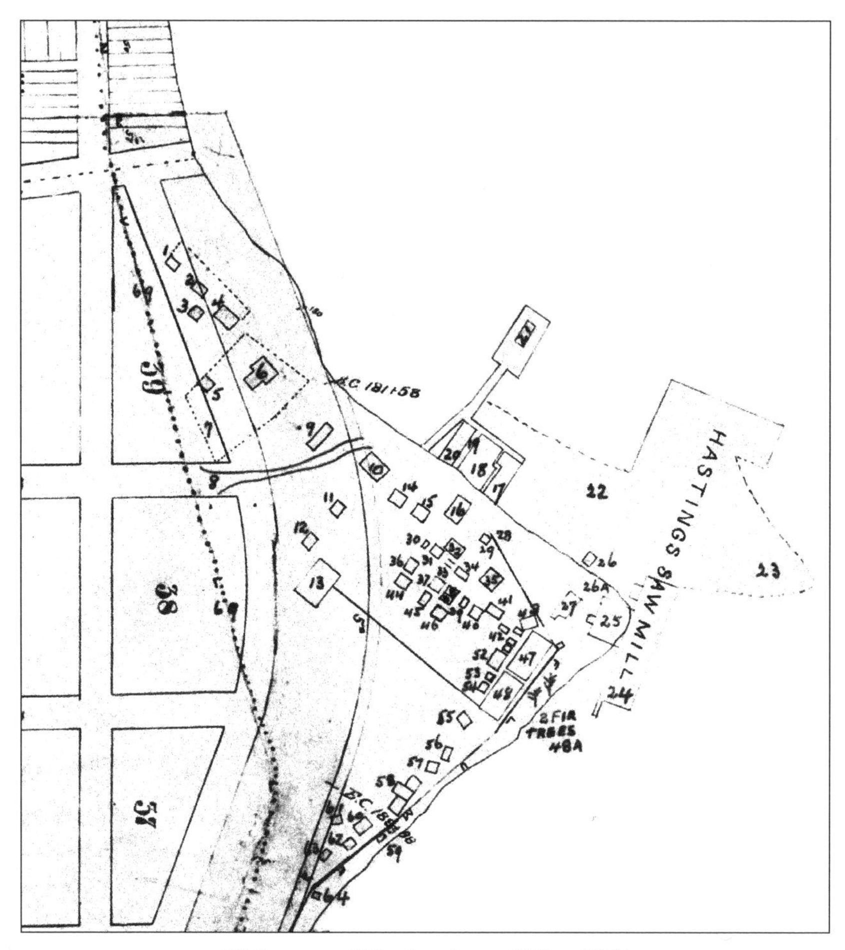

CPR survey of Hastings Sawmill Site, 1886.

and comical dress, who pulled the "Telegraph" about town in ceremonial parades. The Dark Town Fire Brigade has raised much laughter and charitable donations with their hilarious antics over the years, although as a working feature of Hastings Mill firefighting apparatus, the "Telegraph" has long been deemed "unserviceable."

Emma Alexander bustles about the mill grounds in her Sunday finery. Her elegant, midday dinner with guests aboard the *Southern Cross*, has been interrupted by cries of "fire!" As the wife of mill manager Richard Alexander, Emma has established a local reputation for her sweet singing voice—and a headstrong nature not to be trifled with. On June 9, 1862, the beautiful Emma Tammage, one of fifty-nine eligible young women aboard the steam-assisted barque *Tynemouth*, set sail from Dartmouth, England, for Esquimalt, British Columbia. *Tynemouth* was one of the infamous Bride Ships—vessels specially commissioned by English social reformers and religious leaders to transport consenting, unmarried women to the remote wilds and lonely bachelors of the Canadian west coast. The voyage would prove to be one of terror. Storms raged in the North Atlantic, the women experienced severe bouts of seasickness and two violent episodes of mutiny resulted in the most able-bodied sailors being confined to the brig. By the time Emma had arrived on B.C.'s colonial shores, met and married Richard Alexander, and given birth to four children, she had developed a stalwart character seemingly impervious to hardship.

With her husband across the inlet on business, Emma is determined to save the two-storey mill manager's house and estate that she, Richard and the children inherited upon the death of the previous mill manager, Captain James Raymur. While attempting to round up several goats from the spacious grounds and shepherd them to the water's edge, Emma encounters Hugh Campbell. As he proceeds to unload his dynamite onto the mill dock, she rises to the defence. "You can't put these here!"[29] The exhausted Campbell is in no mood to argue. He tersely suggests that the crates be thrown into the water if the fire gets too close. With Emma Alexander's arguments ringing in his ears, he turns back towards town.

CPR architect Thomas Sorby, carrying a small valise stuffed with papers and valuables runs up the Hastings Mill wharf to where the *Robert Dunsmuir*, a small steamer, is anchored. Crowds of people are leaping aboard from every direction—children sobbing with fright, clutching their little treasures, which range from dolls to canaries in cages to incessantly barking dogs. The captain of the *Robert Dunsmuir* is desperately trying to fire up enough steam for the vessel to pull away, acknowledging in the process, that forty kegs of gunpowder are stowed under lock and key below deck. Several men attack the hatch doors with axes as calls to throw the explosives overboard grow loud and shrill. After several anxious moments, enough steam is generated to enable the *Robert Dunsmuir* to begin chugging its way across the Inlet to Moodyville.

Realizing the futility of attempting to save any more household belongings, Jonathan Miller urges his sons to follow their mother and siblings to Hastings Mill as fast as they can. All the while, the city post office, located midway down Carrall Street, has been weighing heavily on his thoughts—along with a famous old saying, "The mail must go through!" Mail in a frontier community like Vancouver is a lifeline. It brings news from families and friends in distant homelands, supplementary cash, transactions relating to business, and often the most welcome communication of all, words of encouragement. When the legendary American "Pony Express" mail service was established in April of 1860, its aim was to ensure delivery of mail between St. Joseph, Missouri, and Sacramento, California—a distance of 1,966 miles—in under ten days. "The Mail Must Go Through"—the famous Pony Express motto[30]—continued to endure in the postal industry north and south of the border, even though the Pony

Express had long been replaced by the telegraph. Jonathan Miller makes a run for Vancouver's Carrall Street post office. Moments later, he is on his way to Hastings Mill, straining under the load of two cast iron boxes stuffed full of registered letters, packages, bills and postage stamps.

On the north side of Burrard Inlet, Frank Hart and Jack McGregor stop in their tracks as they prepare to trundle a large can of fresh milk down to the waterfront. Grimly, the men stare at the Vancouver cityscape, going up in flames before their eyes. As owner/proprietor of Hart's Furniture on Powell Street, Frank Hart has just begun to enjoy a lucrative return by producing fine woodcraft furnishings. Jack McGregor is part-owner of a grocery store on Water Street, not far from the Sunnyside Hotel. The regular boat journey across the inlet to buy milk from Tom Turner's ranch is a welcome diversion for both men from the ever-increasing stresses and demands of commerce in the city.

Frank W. Hart

After a few dumbstruck moments, McGregor states the obvious. "I think we better go back!"

"You cannot!" Hart replies. "The wind is too high, we can't pull back!"

A small crowd is gathering on shore, among them, a young but sturdy native boy from the rancherie. The two men formulate an idea and approach him. "Will you go with us and help pull?"[31] Like most boys of his age group, the lad shrugs off any thought of danger and willingly nods his agreement. The trio cast off into buffeting waves. It soon becomes clear that the vessel's small sail will not hold up against the fierce wind. In a desperate effort to avoid being swamped, Frank Hart and Jack McGregor flatten themselves against the floor of the boat. Somewhat chagrined, they watch as their native companion calmly positions himself at the bow and continues to paddle forward.

Across the mouth of Coal Harbour, lies the great forested peninsula of the military reserve. The "Reserve" is virtually an island, owing to the mudflats which fill with seawater and connect Coal Harbour with English Bay at high tide. Several residents from the reserve's native communities of Chaythoos and Whoi Whoi have gathered on the boulder-strewn beach known as Brockton Point. Word has spread quickly that great smoke is rising from the city of Vancouver. Nineteen-year-old August Jack Khatsahlano stands alongside his family, watching intently as fire consumes the city with astonishing speed. Initial fears of the flames spreading to the large tract of uncleared forest west of Vancouver and breaching the tidal channel are soon dispelled. With the wind blowing smoke and flames northeast towards the mountains, it is clear that the structures of Chaythoos, Whoi Whoi and Brockton Point, as well as the Coal Harbour Hawaiian community of Kanaka Ranch will be spared. Every boat available is being pressed into service as the relief effort mounts.

Out on Burrard Inlet, natives from the Squamish village of Estlahn plunge their paddles deep and surge forward, maneuvering their cedar dugouts towards the burning city amidst the growing flotilla of rescue craft. Many dugout occupants are singing as they approach. A long-standing tradition among the Squamish people, the Paddle Song[32] is sung as travellers near an encampment or village by water. Under happier circumstances, the purpose of the song is to alert people on shore that "visitors are approaching—prepare food and words of welcome." A different version of the Paddle Song is sung when Catholic missionaries are being transported between the villages—a song of faith, comfort and praying to "Mali," the mother of Jesus, in times of trouble. The Paddle Song composed for the missionaries is now being sung as a message of hope to the many people calling for help in the smoke-filled air and churning waters of Burrard Inlet.

High up on the forested slopes to the south of False Creek, a rousing argument has begun. For weeks, there has been much debate about the speed of two highly touted racehorses—one belonging to the town butcher, George Black, the other owned by Sam Brighouse. Brighouse is already something of a local legend, being one of the "Three Greenhorns,"—three Englishmen who had pre-empted 550 acres of thickly forested land to the west of the future Burrard Street in 1862. A large crowd has gathered on either side of the New Road to New Westminster to witness a series of heats between the two men astride their animals. The half-mile straight, completely cleared of stumps and debris, is a Vancouver rarity and the ideal location to hold such an event. Wagers and speculation fly thick and fast. After the initial race, a dispute arises and before the delighted spectators, George Black and Sam Brighouse appear to be ready to come to blows.

Lewis Carter stands amongst the crowd, watching the scene with great enjoyment. Carter, formerly a surveyor with the CPR, has recently opened his own hotel at the southwest corner of Water and Cambie streets. The construction of Carter House has been long and arduous. Carter is immensely proud of the fact that he has cleared the ground single-handedly, a feat which included taking down a massive tree with a twelve-foot girth. On this hot June day, he is enjoying a well-earned break from his labours, or as he puts it, "to get away from the smell of smoke and the cinders and to get a little fresh air."

In the stillness of the afternoon, above the drone of flies and rustle of tree branches, uncharacteristic noises can be heard coming from the direction of town—a series of muffled explosions, accompanied by a steady, metallic clang and the incessant blowing of the Hastings Mill steam whistle. Gradually, conversation ceases as others became aware of the sounds. Black and Brighouse lower their fists. Something is clearly wrong down in the city.

Along with the rest of the crowd, Lewis Carter begins to run back down the hillside. Above the distant treetops, an incredible sight comes into view—a giant column of black smoke with an enormous mushroom-shaped cloud atop, billowing over the landscape. The smoke column spirals into the air like an enormous funnel, with all the outward appearance of a tornado. Carter stares, mesmerized, as flames race to the tops of giant cedars and Douglas firs, then blaze high into the air. "It was a wonderful sight," he would reminisce in later years, "and I don't want to witness another one like it."[33]

Joseph Magnus is among the mourners attending the Masonic funeral for CPR contractor Owen McCormack in New Westminster. Twelve miles southeast of Vancouver, the bustling former provincial capital sprawls along the Fraser River shoreline—a major shipping destination and supply point with well-established sawmills, dockyards and canneries. Adjacent to the river, Columbia and Front streets are lined with businesses. From here, the community banks steeply up a hillside of three hundred feet or more, effectively blocking any views of the mountain peaks on the north side of the Fraser Valley. Magnus and his colleagues have been whiling away the hot afternoon in a local saloon. Between sips of a tall drink, he is among the first to notice the peculiar orange glow and great pall of smoke rising high in the western sky. "Vancouver must be on fire!" he remarks jokingly. Everyone smirks at the comment, although not without a degree of concern. The smoke is certainly indicative of a forest fire and an out-of-control one at that. The men agree that a wise course of action would be to make further inquiries via telephone.

Telephone service between New Westminster and Port Moody has existed for several years and a branch line has recently been extended to Vancouver along the New Road. The New Westminster and Burrard Inlet Telephone Company has thirty-five subscribers and it does not take Magnus and his friends long to find a location to put through a call. Vancouver's one and only telephone exchange is installed at Tilley's Books and Stationery, on busy Carrall Street. Harry Edwards is regularly on duty as the telephone operator, but if the New Road line is down, it is Harry's job to fix it. With twelve miles of line to maintain, service is frequently disrupted. Joseph Magnus breathes a sigh of relief when he hears a voice at the Vancouver end. Relief quickly turns to shock as the men's worst fears are confirmed—Vancouver is indeed, on fire. The voice on the end of the line is distracted and shrill. With a cryptic "Good-bye boys! It's getting too hot to stay here any longer!"[34] the line silences.

Owner/manager of Tilley's Books and Stationery, Seth Tilley, with the help of his son Charley, struggles to salvage whatever he can of his flammable inventory. With flames licking at the outside walls of the store, Seth yanks out the cables connected to his prized Gilliland telephone exchange. Lugging the exchange between them, Seth and Charley stagger down Carrall Street toward False Creek. Their load is about the size of a large medicine chest, but twice as heavy. Before long, Seth is stumbling under its weight, his breath coming in short gasps. At age fourteen, Charley Tilley is developing into a strong and capable young man, full of vigor and optimism.

"This way, father!"[35] he repeats encouragingly, as they make their way.

In New Westminster, the Front Street Commission House of J.R. Homer and Son becomes the base of operations as a massive relief effort is galvanized. Up and down the streets, members of the Hyack Fire Brigade[36] gallop from house to house astride their fastest horses, pounding on doors and spreading the alarm. One glance to the west provides all the necessary proof for anyone harbouring a shadow of doubt about the urgency of the situation. Kitchens throughout the city spring into production. Pantry doors are thrown open, wood stoves fired to life and iceboxes emptied. Before long, a small mountain of food, blankets and clothing lie waiting to be loaded aboard three hastily-commandeered express wagons.

City alderman Thomas Dunn, owner of a hardware store in the Ferguson Block, is one of the many Vancouver businessmen who have travelled to New Westminster for the funeral of contractor McCormack. After the service, he is just sitting down to a late lunch in the Colonial Hotel when Joe Armstrong taps him on the shoulder with stunning news. "You better make haste and get home as there is a chance of your store being burned. Vancouver is on fire."[37] Horrified, Dunn races for the livery stable to hitch up his two-horse team and carriage. He has developed a thriving hardware business over the past few months, but so much more is on his mind. He has a wife and children back in the city.

■ 2:15 P.M.

With explosive fury, the fire sweeps through Vancouver's city core, devouring every building in its path. The office of the *Vancouver Daily News* bursts into flame. The new printing press melts and contorts into a heap of slag. Jars of ink boil and burst. Piles of fresh newsprint, unsold back issues, files, notes and layouts are consumed instantaneously. No amount of residual dampness will save the wooden wash tubs of Wah Chong's laundry. Glass washboards shatter, while wooden laundry agitators and baskets full of wooden clothes pegs frazzle and burn. Gassy Jack's landmark Deighton House—Vancouver's first-ever elite accommodation, is fully ablaze—decorative wallpaper, fixtures and feather bedspreads accumulated with so much pride over the years, being destroyed by voracious flames with a seemingly reckless abandon. A huge tongue of flame leaps from the hotel's rooftop across Maple Tree Square, igniting Sciutto's Bakery on the opposite corner. The magnificent old maple tree flares like a giant beacon, its parched green leaves abruptly withering into ash. Businesses occupying the Ferguson Block building are swept over. The first strawberries of the season sizzle and burn in their boxes at James Hartney's grocery store. Nails fuse together in their unopened bags at Thomas Dunn's Hardware. Desks, files and chairs in the upstairs CPR offices are consumed.

Racing up Powell Street, Captain Jackman notices Joanna Hartney, wife of storekeeper James Hartney, desperately struggling to climb out of the Oppenheimer brothers' root cellar. He hoists the terrified woman out and together they run for Hastings Mill. Seeing no other means of escape, Jackson Abray and Mr. Anderson have also chosen to jump into the cellar, hoping that the wide cement expanse will give them some degree of protection. Crouching low on the floor, Abray glances at the box of lead type he has placed nearby. The letters are contorting as the lead begins to melt in the intense heat.

"We've got to get out of here!"[38] he tells his companion grimly.

Clambering out of the cellar, the men begin a southward run, hoping to reach the shore of False Creek. Barely underway, they cross paths with Jonathan Miller. Gasping for breath, the postmaster has been forced to abandon the iron boxes of mail, although he is still clutching a small metal cashbox.

"There's no use trying to get out this direction!" Miller informs them flatly.

"There's no chance to get out any other way! Everything is on fire!"

The postmaster stubbornly shakes his head. "Yes, we can get out! I know the old logging roads and we can get through them even if there is a fire!"[39] With handkerchiefs clenched tightly over their noses and mouths, the men run east for Hastings Mill through an interweaving maze of logging trails, little known by most Vancouver residents, but well memorized over the years by Jonathan.

At a large clearing near the eastern edge of town, Dr. McGuigan and John Blake come across an abandoned wagon. Knowing that they can't continue their pace encumbered with heavy loads, they decide to stash their rescued belongings under the wagon.

In the same moment, John Blake suddenly announces that he has forgotten his pipe. Dr. McGuigan carries on running, failing at first to notice that his friend has turned back towards town, determined to retrieve the missing pipe. Upon realizing that he is alone, McGuigan backtracks, quick to voice his anger and incredulity. "Supposing your girl was shut up in your office and the stairs were in flames. I would risk my own life and go with you myself, but for a pipe and tobacco, it is nonsense, Blake!"[40]

John Joseph Blake

Blake will not be persuaded to abandon his pursuit of the errant pipe and Dr. McGuigan can only stare, open-mouthed, as he disappears off into a thick veil of grey haze and flying embers. Feeling compelled to wait, his thoughts revert to his younger brother Thomas. How is Thomas faring in all of this? Short minutes later John Blake is back, hair badly singed, pipe firmly clenched between his teeth. With the doctor shaking his head in disbelief, the men race on towards Hastings Mill.

Uncertainty nags at Charles Gardner Johnson as he continues to fight the flames inside the Balfour Hotel. How much further has the fire advanced since they entered the Balfour? A quick check out the front door confirms an awful truth—flames are encroaching on all four sides of the building. Racing back to his companions, Gardner Johnson shouts that they must leave immediately. With soaking wet handkerchiefs clamped over their mouths, the men run outside, but not before a grateful George Bailey quickly stuffs a bottle of brandy into Charles Gardner Johnson's pocket.

Flames are everywhere, rapidly devouring every building, every tree, every clump of surrounding brush. Mere footsteps from the Balfour, a panic-stricken man babbling unintelligibly, throws bucketfuls of water onto a woodpile. Glassy-eyed with terror, he flatly refuses all urgings to abandon his efforts and run. In frustration, the four men press on without him, fighting their way southward through thick smoke and flying embers. A nearby house collapses, sending showers of flaming debris towards them. Not far ahead, Gardner Johnson notices a shallow, gravelly hollow in the earth, left from the removal of a tree stump. With no flammable brush in its immediate vicinity, the hollow seems as good a place as any to seek refuge—a long shot, but quite possibly their only hope in an increasingly hopeless situation. The men throw themselves down, side by side. In the same moment, Gardner Johnson notices a dark, rectangular-shaped object lying nearby. Crawling over, he discovers that it is a valise—relatively new-looking and presumably abandoned by its owner. Thinking that the valise might offer the men some protection from flying sparks, he lugs it over and positions it in front of his companion's heads.

Huddling together, the men bury their faces in the gravelly earth. Charles Gardner Johnson can feel the ever-increasing heat through the soles of his boots. The bottle of brandy shatters in his pocket. John Boultbee swats frantically as an ember sprouts into a small orange flame on his shirtsleeve.

"I have to get out!" George Bailey suddenly announces. "I can't stand this anymore!"

The others are unanimous in their response. "Lie still, George!"

Bailey grows more and more agitated. Despite their best efforts, Gardner Johnson, Boultbee and Samuel Macey cannot subdue his panic. Moments later, they witness the horrifying spectacle of George Bailey, running in maddened circles with heartrending cries, burning to death not twenty feet away. Lying stock still, Gardner Johnson grasps Boultbee's hands in his own.

"John, if I live and you don't, I will look after Mrs. Boultbee and the family and do all I can for them."

"Charlie, old man," John Boultbee replies softly, "if I come out of this all right and you don't, I will do all in my power for your wife and child." As the men lie in the gravel clutching each other, a loud popping sound rings in their ears directly over their heads.

"What can that be?" Gardner Johnson asks.

Up until this time, Samuel Macey has been lying silently alongside the two others. Suddenly he speaks up.

"Say pard, can you pray?"

"You should pray for yourself!" Gardner Johnson replies, with just a tinge of annoyance.

The popping sound continues. As if he did not already have enough pain to contend with, John Boultbee winces as several tiny blows ricochet off the back his skull.

"There are cartridges in that valise!" Macey tells them. "The heat is making them explode!"[41]

If the situation were not so grim, Charles Gardner Johnson imagines that he would laugh. Here they are, surrounded by flames, and now at the mercy of an exploding valise. He burrows his face deeper into the ground and presses his hands to his ears.

Father Clinton hauls relentlessly on the bell rope of St. James Church. Mere feet away, flaming joists are crashing to the floor, setting alight the tidy rows of wooden benches in the sanctuary. Suddenly, wet hands grip his shoulders. William Gallagher has waded to shore and entered the church. He gently, but firmly suggests that it is time to leave. Father Clinton stubbornly maintains his hold on the rope, steadfast in his belief that more lives will be spared if he continues to sound the alarm. Only as the bell turret begins to blaze and panels of stained glass shatter in their arched window frames, does he submit to the strong arms pulling him away. Gallagher steers the priest down a short pathway behind the church to the Burrard Inlet shoreline.

The scene in the water can be likened to a chaotic mass baptism. Scores of people are crouched low amidst the waves, struggling to maintain their footing. Father Clinton wades out amongst them, quickly gathering a young boy into his arms while bidding another to clamber aboard his shoulders. Calling again and again for calm, he watches as his beloved little St. James Church crashes to the ground in flames.

The fierce wind has churned Burrard Inlet into a maelstrom of whitecaps. Anchored in the middle of Coal Harbour, the ninety-foot barque *Robert Kerr* sways and groans protestingly, anchor chain stretched taught. The *Robert Kerr* is by far the largest vessel in the harbour, though not by any means the most seaworthy. The previous summer, en route from Victoria to Burrard Inlet, on the final leg of a long and grievous journey from Liverpool, the *Robert Kerr* had run aground off San Juan Island and sustained severe damage. Crewmen had managed to coax the ship to the Hastings Mill dock under tow, where bilge pumping and minor repairs took place—but inspectors quickly declared that the *Robert Kerr*'s long haul, ocean-going days were over. The barque was put up for auction and purchased by Hastings Mill stevedore, Captain William Soule. Derelict and forlorn, the *Robert Kerr* has remained anchored in the harbour ever since—no one quite sure of what to do with her.

Succumbing to the relentless gale, the *Robert Kerr* is blown steadily eastward, her anchor gouging a deep swath across the murky bottom of Burrard Inlet. Not far off the barque's port bow, several vessels approach—steamers and tugs, plus a ragtag assortment of virtually anything else that can float. Vancouver citizens fleeing the fire have grabbed hold of driftwood logs, scrap lumber, wooden barrels, even the very planks of hotel wharves. Many of the hardiest individuals, some with little more than a barrel length of wood beneath them, paddle furiously for the *Robert Kerr*.

Robert Kerr *at anchor in Burrard Inlet, 1886.*

Captain Dyer, retired as a sea captain and now the *Robert Kerr*'s on-duty watchman, is back aboard the vessel in his charge, already having ferried one group from the burning city wharf out to the *Robert Kerr* aboard a small rowboat. Now, more and more vessels—both conventional and makeshift—are converging. Captain Dyer watches with growing apprehension as the fleet approaches. His misgivings turn to outright concern as one overloaded craft after another pulls up and small swarms of passengers begin clambering up the side of the *Robert Kerr*'s hull like marauding pirates. Men stand precariously atop each other's shoulders to heave themselves over the railing. The steamer *Senator* pulls up, with Emily and Alexander Strathie among its passengers. Shouting for calm and order, Captain Dyer finds himself flattened against the pilothouse wall as a steady stream of bedraggled fire refugees surge aboard—women clutching sodden skirts and crying children, elderly folk barely able to climb the rope ladders that have been quickly tossed over the side. In subdued shock, everyone gathers side by side on deck, gaping in awe as a solid line of fire consumes their homes and businesses on shore. Standing among them, Mayor Malcolm MacLean unabashedly blinks back tears, having arrived by dugout canoe. Businessman John Rankin stands alongside him, transfixed by the spectacle of Vancouver's destruction.

"Like waves on an ocean!"[42] he grimly muses to himself.

■ 2:30 P.M.

Mr. Anderson falls to the ground near Gore Avenue.

"I can't go any further!" he gasps.

Jackson Abray is not to be deterred. He hoists the man up in his arms and stumbles along in choking smoke and searing heat.

Mr. Anderson is flailing helplessly, but somehow maintaining a grip on his valise. He tries to run on his own but falls a second time.

"Go on, Jack,"[43] he says simply.

It has become a desperate, every-man-for-himself situation. Feeling sick at heart, Abray does as he is told, and continues to make his way to Hastings Mill—tears not altogether caused by the smoke streaming into his handkerchief.

Alderman Charles Coldwell is determined to save his home. The two-storey Coldwell house is the westernmost of four residences that occupy the perimeter of the Hastings Mill townsite closest to Vancouver—a small oasis of high society amidst the largely ramshackle collection of cottages and shacks housing mill employees and their families. Alderman Coldwell's nearby neighbours include Reverend Thompson, mill stevedore Captain William Soule and his wife Theresa, and mill manager Richard Alexander, his wife Emma and four children. The estate-like Alexander property, by far the largest in the townsite, includes a barn and fenced corral. On any average day, Hastings Mill's blueblood neighbourhood is a picture of serenity. There is no serenity now as Charles, perched precariously on his rooftop, swats at sparks with a wet blanket, every now and again firing shots into the air with his revolver. "Let 'er go, Gallagher!"[44] he yells maniacally.

Alderman Charles Coldwell

Frank Hart, Jack McGregor and their native companion, having managed the crossing from Moodyville, pull up on shore by the mill's sawdust sports field and make a run for town—their sole objective to reach their respective businesses. They quickly realize the impossibility of it.

"I believe the town is gone!" McGregor states simply.

The two men turn their efforts towards helping Reverend Thompson, who is attempting to save as much furniture as possible from his home. Frank Hart carries out a piano stool, much to his frustration damaging it in the process. In the same moment, he looks up to the rooftop of the neighbouring Coldwell house as a crack of gunfire pierces his ear.

"Charlie Coldwell, what the heck are you doing up there, you damned fool!"[45]

Alderman Charles Coldwell, somehow appearing more jubilant than frightened, yells back that he is trying to create an updraft, in hopes that it might change the wind direction. He sits upon the rooftop, calmly firing off one round after another, for all intents and purposes looking as if he is on a morning duck hunt in the False Creek marshes. The men hard at work below shake their heads. It appears that Alderman Coldwell has well and truly lost his mind.

Whether or not Coldwell's unorthodox methods of firefighting have succeeded is open to conjecture, but within the next few minutes, the wind actually begins to alter course. As the flames that have consumed Vancouver so voraciously reach the waters of Burrard Inlet and find little else to feed their ravenous advance, they quickly lose their momentum. As onlookers crouching low in the waves offshore wait and watch, the choking blast of smoke drifts into lazy eddies, parting now and then to reveal a landscape of destruction on shore. Wet blankets continue to be re-soaked and re-swatted, but now with a mood of optimism as opposed to futility. A bucket brigade slows. Ladders are hauled down.

Up on the rooftop of his undamaged house, Alderman Charles Coldwell lowers his revolver, with the barest hint of satisfaction on his face.

Hundreds of people stream out of town across Westminster Bridge, the single crossing between the north and south shores of False Creek. Lewis Carter, lips pursed in determination, makes his way in the opposite direction, alternately running and walking. He passes men with singed hair and faces streaked with soot. Exhausted, crying children are being dragged along, clinging to their mother's ash-smeared Sunday dresses. Smoke continues to billow overhead, obscuring the sun and casting the landscape into a subdued, copper-coloured glare.

Just after crossing the bridge, Carter encounters his first graphic evidence of the fire's ferocity. A light wagon passes, the driver weaving his way down Westminster Road as well as he is able. Something covered by a sheet has been loaded aboard. As Carter steps up on a log to allow the wagon to pass, a swirling gust of wind blows off part of the sheet revealing the body of a woman. Carter recognizes her. She recently arrived in Vancouver—a large woman, weighing perhaps over two hundred pounds, a weight that must have hampered her escape. In horror, Carter sees that one entire side of her body is burned away. Her hair and eyebrows are gone, and one arm is a charred fragment. One young woman passing nearby who sees the ghastly sight utters a shriek of horror, and falls to the ground. Carter runs to the edge of False Creek, fills his hat with water and dashes it upon her face. Wringing her hands, the woman comes to her senses momentarily, only to faint again. Leaving her to the care of others, Carter continues on his way, receiving many looks of incredulity as he hurries in the direction of town.

"Turn around, or you'll be burned for sure!" a man calls to him.

"What do you know of Carter House?" Carter asks anxiously, ignoring the warning.

"Carter House?" the man replies. "I believe it survived. The fire spread to either side of it."[46]

With this promising report, Lewis Carter becomes more anxious than ever to reach his hotel.

■ 3 P.M.

An eerie calm settles over the landscape where Vancouver once stood. Spot fires continue to burn as dying flames consume the last fallen timbers. With a final groan, the skeletal remains of the Sunnyside Hotel collapse into a steaming heap. The creosote-coated pilings beneath the new addition ignite with ease, bringing down the entire structure like a giant house of playing cards. The old maple tree is no more, its once glorious green canopy burned to the bare earth. By contrast, the forested regions to the east and west of the city are as lush and green as ever. The fire took dead aim where it could do the most damage. Remarkably, the Regina Hotel stands tall and unscathed in its Cambie Street clearing. A small group of men, exhausted from their long rooftop battle, unabashedly help themselves to the hotel's liquor stocks.

Back in 1868, when Captain Edward Stamp wanted a sturdy new building of durable Douglas fir to serve as a supply base for his fledgling sawmill, Hastings Mill Store was constructed. Standing remarkably sound and unscathed atop its wooden pilings, it is now quickly commandeered as a makeshift hospital and base-of-operations. A chorus of hysterical cries pierces the air as families and friends seek to reunite. Shelves of groceries line one side of the store. Clothing, hardware and other household goods take up the opposite side. In between, victims of all ages lie sprawled on the floor—some moaning in agony, others staring blankly in silent shock. Several soaking wet individuals cluster about the mill stove—an oil drum lying on its side in a bed of sand and bricks. A badly burned woman is discreetly carried to a quiet shack on the Mill grounds and placed upon a box mattress, wrapped in a blanket.

Dr. McGuigan, having made it to the mill with John Blake, soon finds himself overwhelmed with work. He moves quickly among burn victims, dispensing the small amount of opium he managed to save from his surgery to those in the most severe pain. With the help of Emma Alexander, he applies soothing salves and dressings to scorched hands and blistered feet. Many individuals have had to run over hot coals during their escape. Eyes stinging from smoke and heat are gently washed. As more and more patients converge, supplies rapidly diminish and runners are sent to pick as many skunk cabbage leaves as can be found from neighbouring marshlands. Used together with scraps of lint soaked in carbolated oil, the soft and pliable skunk cabbage makes an effective poultice. The wooden flume carrying fresh water from Trout Lake is choked with ash making drinking water to quench parched throats in short supply.

The fire has proved to be fickle at the Hastings Mill townsite. Alderman Charles Coldwell's house stands undamaged, as does the stately Alexander residence, although the latter has lost its neighbouring fence and barn. The Soule and Thompson houses are completely destroyed. A large tree has blown down across Alexander Street, blocking off vehicle passage.

Many people are wet and cold, having taken to the waters of Burrard Inlet or False Creek to escape the flames. The north shore community of Moodyville opens its heart to Vancouver's plight. As one rescue craft after another pulls up dockside, fire refugees are gently and sympathetically escorted to welcoming Moodyville homes, provided with ill-fitting—albeit dry clothing and soothing cups of tea. Since its 1862 beginnings as site of the first sawmill on Burrard Inlet, Moodyville, like Vancouver, has grown into a thriving community.

Hastings Mill townsite, July 1886. The two-storey white building (background right) is the Alexander residence.

Sunday, June 13, 1886

When Jonathan Miller's family arrives at the Moodyville dock aboard one of many converging rescue steamers, no one is more relieved to see them than members of the Patterson household. John Patterson, his wife Emily and four children were early residents at Hastings Mill, before moving across the inlet to Moodyville in 1874. In a time when playmates were few amidst the sawdust and industrial clamour of Burrard Inlet's frontier settlements, the Miller and Patterson children had enjoyed each other's companionship immensely. Emily Patterson assesses each fire refugee carefully for any signs of shock or injury. Emily has come to be regarded as something of a saintly individual over the years—providing nursing care to residents up and down the shores of Burrard Inlet, despite her complete lack of formal training. Twenty-two-year-old Alice Patterson shares her mother's deep sense of care and compassion, escorting her emotionally and physically exhausted friends to the comfortable Patterson home.

A distraught and exhausted Mrs. Miller remains at the Moodyville dock clutching a rescued prayer book, watching for her husband. When Jonathan Miller finally steps ashore, he is carefully balancing his spectacles atop the post office cash box.

"I saved my glasses, Mrs. Miller!" he says with a feeble smile.

The Miller children babble on excitedly, while Alice listens with a sympathetic ear.

"What did you think about it all?" she asks Carrie Miller.

"My only thought was to get Ma out of the fire," Carrie replies matter-of-factly.

"And what about you?" Alice continues, turning to young Alice Miller.

The little girl crinkles up her nose in disgust.

"Thank goodness that old coat I hate got burned up!"[47]

Thomas Dunn has been galloping along the New Road to Vancouver as fast as he can urge on his team, desperately trying to sort fact from rumour in the course of his journey. Arriving at the Gladstone Inn, a roadhouse well en route, he had found several people milling around and anxiously made inquiries about the fire. A man by the name of Wilson shrugged and replied that it was "not very much, only behind the town." Dunn eased up on the team, but had not gone another five hundred yards before meeting up with John Gillis with his wife and family aboard a wagon loaded with household goods. Everyone appeared to be in a state of great dismay and Dunn soon received an entirely different report—"the town was all burned down."[48]

Descending the hill towards Westminster Bridge, Dunn passes more and more wagons being driven away from the city—some lurching precariously with the added weight of hastily thrown-aboard luggage. Reaching the bridge, he is confronted with a dreadful sight—a smoke-filled sky and bright flames consuming the last skeletal remains of Vancouver homes and businesses. It is impossible to see the extent of the destruction beyond the thick haze enveloping the city. Leaping out of the carriage, he breaks into a run, asking amongst the traumatized passersby if anyone had word of his wife and children. Someone replies that they think one of his boys had been burned. Many individuals flatly advise Thomas Dunn not to go any further—the route is quite simply, impassable.

■ 5 P.M.

With many patches of the ground still scorching hot and smoldering timbers lying across roadways and paths, individuals attempting to traverse the ruined landscape must choose their route with care. The Reid family, caked from head to toe with mud but grateful to be alive, gaze with shock and awe at the damage surrounding the sanctuary of their ditch. Their house has been consumed, fallen timbers still too hot to probe through for anything possibly salvageable. Duncan Reid's coat and hat have burned off him. Great holes, the size of dishpans, have burned into the blankets that covered Christina Reid and her children. Duncan hoists up baby Minnie in his arms and begins leading Christina—resolutely still clutching her six silver teaspoons and parasol in addition to Jemima's hand—on a painstakingly slow journey in the direction of Hastings Mill.

Thomas McGuigan arrives at Hastings Mill, anxious to learn if the city archival records that he left aboard a scow in Burrard Inlet have survived. Finding a way back from False Creek through a wilderness of burned wreckage has been no simple endeavour for the city clerk. Like many others, he has slogged through the shallows of False Creek's most northerly arm and made his way through a four-block stretch of scorched earth, strewn with debris along Columbia Street. Years ago, the going might have been much easier—or at the very least, cooler underfoot. Columbia Street, and much of the surrounding area, was formerly bogland—a tidal marsh that virtually connected False Creek with Burrard Inlet. Much of the bog was now filled in and what might have served as both a firebreak and a lifesaving escape route was no more. Still feeling the ground's heat beneath the soles of his boots, Thomas McGuigan wanders among the distraught crowd.

Mayor MacLean, having been rowed from the *Robert Kerr* to the dock of Hastings Mill, is besieged with distraught individuals —most of whom have lost everything but the clothing on their backs. Making his way through the crowd with assurances of help, he comes across McGuigan. Relieved to find his city clerk alive and able-bodied, the mayor quickly obtains a sheet of paper and ponders over the right wording for a telegram. He has already decided that it must go straight to the country's highest authority—the office of the prime minister. It must somehow convey the absolute urgency of Vancouver's situation. Bad fire in Vancouver? Fire destroys Vancouver? Ashes...everywhere there are ashes, as far as the eye can see. Vancouver *is* ashes. Mayor MacLean gives Thomas McGuigan a carefully-hand written dispatch:

> To Sir John A. Macdonald. Our city is ashes. Three thousand people homeless. Can you send us any government aid?[49]

TELEGRAM.

To Sir John A McDonald 307
301
pd

From Vancouver B.C. 14 Ottawa, 14 June 1886

Our City is ashes three thousand people homeless can you send us any government aid

M. A. MacLean
Mayor

Mayor Malcolm MacLean's telegram.

A similar message is designated for Mayor William Howland of Toronto:

> Vancouver is ashes; 3000 people homeless. Please send aid at once. Please repeat this message to other Canadian cities.[50]

The mayor's instructions to McGuigan are brief. Hire a fast horse, get to New Westminster and have the messages cabled to Ottawa and Toronto. Minutes later, the city clerk is galloping down the New Road.

Another galloper arrives at Hastings Mill from the opposite direction. A man leaps off his sweat-lathered horse and announces that he has a message for the citizens of Vancouver from New Westminster. Mayor MacLean hurries over to hear the welcome news that three wagonloads of relief supplies are being readied to travel to the stricken city via the New Road. The mayor is impressed that New Westminster has received word so quickly—and relieved to hear that regions to the south and east of Vancouver escaped the fire's onslaught.

Now there are other options to consider. The New Road terminates at Westminster Bridge, whereupon Westminster Road—in later years to be re-named Main Street—continues on into the city. The fallen tree between the mill and what was formerly the central business district of Vancouver, has rendered Alexander Street impassable. No news has arrived, as of yet, on the condition of other thoroughfares within the city, although there is little doubt that they will be impassable for wagons. While many Vancouverites have converged at Hastings Mill, Thomas McGuigan reported that there is also a large crowd at False Creek. Burned-out fire victims are likely scattered over a wide area throughout the city. The most logical course of action will be to have everyone gather at a central location to await the promised supplies. Mayor MacLean issues his first post-fire, citywide order. All able-bodied fire survivors in need of assistance are to make their way to the southern end of Westminster Bridge, to rendezvous with a supply convoy on its way from New Westminster.

Thomas Fisher, having made a hasty departure from Coal Harbour, has found his family alive and well at Hastings Mill. To their great relief, he and Lavinia receive welcome word that Walter and Lavinia Rosella are among survivors aboard the *Robert Kerr*. With the abatement of the driving wind, the old ship has come to rest not far offshore. Aboard his small sailboat, Father Clinton is ferrying displaced family members to joyous reunions on both sides of Burrard Inlet.

Emily Strathie fretfully paces the deck of the *Robert Kerr*. The thick clouds of smoke have parted, and she can see that the fire has dwindled—the fire that she had all but prophesied. For weeks, she and Alexander had felt uneasy. The city was tinder dry and, in the Strathie's minds, "if a fire broke out the place would burn like matchwood."[51] Alongside her, Captain Dyer surveys the scene through a pair of field glasses.

"May I have a look through those?" she asks him. "I'd like to see where I lived."[52]

Emily pans the glasses carefully up and down the shore. The waterfront cityscape that she has known is gone, and there is no sign anywhere of the trunk that Alexander dragged to the beach. Quickly she seeks out her husband and insists that they return to shore by any means possible. Alexander Strathie, in a state of shock and despair that his past several weeks of meticulous carpentry work has been so quickly and completely destroyed, wants no part of the venture. Undeterred, Emily spies a man aboard a small rowboat tied up alongside the *Robert Kerr*, and politely asks if he will row her to shore.

The tide has risen over the past few hours, submerging all but the largest boulders and snags. As Emily and her rower near landfall, she notices two native women stepping carefully through the burned debris. Looking closer, she sees that one of the women is carrying something red and white in her hand—an object Emily quickly recognizes as her mother's pincushion.

"Where did you find this?"[53] she asks, catching up with the women.

One of them points to a location not far off shore. Excitedly, Emily asks her rower if he would mind wading out and reaching down into the murky waters. Obligingly he does so, plunging his arm down and bringing up a thick handful of seaweed. Something small and black is dangling from the mass. Emily stares in amazement. That something is her much-treasured golden locket. Of course its present appearance is far from golden and the links of the chain have melted, but otherwise, the heirloom is fully intact. A second dip proves equally fortuitous, as the rower brings up Alexander's silver watch, now useless as a timepiece but otherwise in remarkable condition. Any remains of the trunk itself are nowhere to be found. The jewelry pieces had evidently settled deep enough in the water to be protected from the worst of the fire's intense heat.

Back aboard the *Robert Kerr*, Emily triumphantly produces her findings to a much-relieved Alexander, who has been beside himself with worry over his wife's whereabouts. The Strathies resolve to keep their treasures, however tarnished, as permanent mementoes of the day Vancouver burned.

Alexander Strathie's pocket watch.

The Boultbee homestead and Gardner Johnson cottage have survived unscathed, being located well to the south of the fire's advance. A few neighbouring houses, the Royal City Planing Mill and Bridge Hotel are also untouched. People are on the move both into and out of the city along Westminster Bridge. The wooden thoroughfare connects the north and south shores of False Creek at a narrow point about halfway between its English Bay confluence and eastern tributaries. Many individuals hurry along with purpose—faces grim, eyes set—while others shuffle, semi-delirious, with no apparent destination in mind.

Minnie Gardner Johnson stares aghast as two blackened figures are helped across the front yard of the Boultbee home. Despite excruciating pain, Charles Gardner Johnson and John Boultbee have managed to crawl back home on their hands and knees from the burned out streets of the city. Their tattered clothing is pockmarked with burn holes and thoroughly coated with soot. Both men collapse at the foot of the porch steps, weeping uncontrollably.

Lewis Carter glumly surveys the blackened ruins of Carter House. He has made an exhausting journey back to his property, following the north shoreline of False Creek through a tangle of underbrush and fallen logs. With the change in the wind direction, smoke from the fire drifted southward and forced him to lie down many times and bury his face in the ground. Now he has discovered that the promising information he had been given was wrong. While the Regina Hotel has been spared, Carter House has burned to the ground. It would have been an easy error to make—two white-washed hotels of similar height, located across the street from each other. All of Lewis Carter's possessions, with the exception of the clothes on his back, are gone. He soon receives word of Mayor MacLean's instructions, and begins making his way back to False Creek.

Exhausted, frustrated and sick with worry, Thomas Dunn has returned to New Westminster. His one consolation at present is that he has provided a small degree of assistance. Henry Abbott and Mr. Terhune have travelled back with him—the CPR superintendent vowing to set the wheels of charitable relief in motion. Since Vancouver cannot be accessed via the New Road, Dunn has decided that he will try using an alternative route with a fresh team of horses. Douglas Road was built in 1865 at government expense to give New Westminster residents an access route to the beachside resort of New Brighton on Burrard Inlet. New Brighton, now called Hastings, is still a good three miles east of Vancouver—but if Dunn is lucky, the road will be passable and he will be able to find his wife and children.

■ 9 P.M.

The sun, a mottled orange orb through the lingering haze of smoke, edges towards the western horizon. Bedraggled survivors pick their way towards Westminster Bridge through an ash-strewn wilderness of wreckage. Silence prevails among them. A guilt-stricken William Gallagher follows along, inwardly wondering if there is anything that he could have done differently back at the roundhouse clearing site to prevent all of this misery from happening. The makeshift bivouac ordered by Mayor MacLean, has already begun to take shape along False Creek's southeast shore. Stout branches have been hastily cut to fashion rudimentary tents and lean-tos. One-Armed John Clough nonchalantly arrives with a thick pile of woolen blankets. It had long been rumoured that he kept a secret stash of comforts somewhere off in the woods for his prison charges. Opinion is unanimous as fire refugees gratefully receive Cough's blankets—the gruff old city jail keeper is truly a softie at heart.

Elsewhere, fire refugees prepare to spend the night wherever they happen to find themselves. George Schetky resolutely plants himself in a chair at Hastings Mill Store, safeguarding his rescued accounting books and $600 in cash. Lauchlan Hamilton and his family gratefully receive a rowboat ride from George Cary to their board-and-batten shack on the south shore of False Creek—a rustic, albeit comfortable abode that they had been using during Lauchlin's survey work in the area. Several individuals curl up on the pungent floorboards of "Spratt's Ark,"[54] a 240-foot steam-driven scow moored at the foot of Burrard Street, used until recently as a plant for rendering herring fish oil. Out on the *Robert Kerr*, every available inch suitable for sleeping accommodation has been taken up—cabins, decks, galley, and even the ship's hold. Captain Soule and his wife, now homeless like so many others, distribute the *Robert Kerr*'s sails to improvise hammocks, curtains and bedding for an estimated three hundred exhausted refugees. Ultimately, there is no spare room for the Soules aboard their own ship, and they gratefully accept accommodation aboard the German barque *Von Moltke*, which has arrived in the harbour.

Over in Moodyville, the hotel is full, and store shelves are being emptied. Moodyville sawmill manager Benjamin Springer tactfully requests that individuals taking shelter at the Mount Herman Hall Masonic Lodge refrain from sitting on the much-revered Masonic chairs. Moodyville residents take great pride in their community—the first west-coast population centre north of San Francisco to have electric lighting. Lumber processed at Moodyville mills can be loaded aboard ships day or night. James Ross will spend a restless night seated against a brightly lit fisherman's shack, his wife and baby daughter sharing a cot within. With his family safe, Ross now focuses on the probable loss of his *Vancouver Daily News* office, and how he will endeavour to record this defining day in Vancouver's history.

Some Vancouverites have chosen to make their way to hotels in New Westminster, among them restaurant owner Sherman Heck. With his business destroyed, Heck telegraphs his mother in Seattle to send "an immediate remittance of seventy-five dollars."[55] Points further up the Fraser Valley are making inquiries about the thick pall of smoke which darkened skies over Mission and dropped fine ash as far distant as Chilliwack.

Armed with lanterns-full of kerosene, Postmaster Jonathan Miller, Thomas Dunn and Thomas Sorby travel across Burrard Inlet from Moodyville for a closer look at the ruins of Vancouver.

Thomas and Isabella Dunn and family.

Thomas Dunn has had an exhausting day of travel, but once he had arrived at Hastings townsite and learned that his family was alive and safe in Moodyville, he found his spirits and resolve soaring. After a brief but joyous reunion with his loved ones, he was determined to learn the fate of his hardware business. Miller is less optimistic, but still anxious to see if anything remains of his post office.

After edging their way across the damaged pier at the foot of Carrall Street, the men gingerly make their way through still-smouldering ruins of the city. A small crowd of ruffians has broached an intact whiskey keg that has washed ashore, and are engaging in an alcohol-fueled rampage through the streets. Hunting for plunder, they recognize and surround the postmaster. Much to Miller's relief, several strong men appear out of nowhere to hold the would-be combatants at bay. The post office is a ruin. The contents of its iron safe—silver and gold coins—have melted together into a heap. Continuing their search, the men discover that two safes appear to be undamaged in the wreckage of the Ferguson Block. The safes are still red hot and cannot be opened, but Thomas Dunn wastes no time in commandeering two men to stand guard over them for the night. Minutes later, Jonathan Miller makes a more grisly discovery in the debris of an old stable at Hastings and Columbia streets—four bodies, huddled together in an almost unrecognizable mass. Little can be done, as evening darkness descends.

■ 11:50 P.M.

From Westminster Bridge, the New Road disappears into inky blackness. Hunger gnaws at the exhausted crowd encamped along the south shore of False Creek as they await the arrival of the promised supply wagons. The light of a nearly full moon supplements the faint glimmer from a small supply of kerosene lanterns, turned down low to save precious fuel. At long last, the welcome sound of hoof beats and creaking of wheels signal that help has arrived. Hard-boiled eggs packed in soda cans and fried-egg sandwiches are quickly distributed to gratefully outreached hands. Children are served first, women next, with men respectfully waiting for anything left. Minutes later, a rowboat glides up to shore. Four exhausted sailors from Port Moody, the easternmost township on Burrard Inlet, clamber out and begin distributing supplies. CPR contractor Andrew Onderdonk has a construction office in Port Moody with a telephone connection to New Westminster. When word came through of the fire, Port Moody residents, like those in New Westminster, had mobilized into action.

To everyone's astonishment, the sailors have made the journey from Port Moody all the way down Burrard Inlet, through the Narrows, across English Bay and up False Creek to the bivouac—a staggering distance for even the most seasoned rowers. A sense of acute embarrassment descends upon the crowd as it is realized that the food from New Westminster has been drastically depleted. There is little left to offer the rowers. Some rummaging amongst empty boxes turns up a small, unopened parcel with an attached note. "This is very little, but all I have." As the hungry sailors divide up the sandwiches found inside, one of them gratefully raises his blistered hands to the east, declaring "God bless the people of New Westminster! May they never suffer such tribulation as that which surrounds us here today!"[56]

Monday, June 14, 1886

■ 4 A.M.

Don MacPherson strides back and forth in the pre-dawn light, busily pulverizing the last remains of timbers strewn over ground where his CPR Hotel once stood. Everything is still hot to the touch and he has had to pour many buckets of Burrard Inlet seawater over the ruins before progressing with his work. Hastings Mill manager Richard Alexander has announced that free lumber will be available for the taking to all fire victims. If Don MacPherson has his way, the CPR Hotel will be the first of many establishments to rise from the ashes within days after the fire.

■ 6 A.M.

Margaret Sanders bustles about her kitchen with Annie Ellen's help, stoking the fire and emptying cupboards. She is grateful beyond measure that her family has survived—Annie Ellen and her brothers had made a fast trip home from Sunday school the afternoon previous, upon the urgings of Reverend Thompson. The Sanders' Prior Street home, one of the few surviving structures on the eastern edge of town, has become an impromptu emergency shelter. Seventeen people doze fitfully on the dining room floor. Another five emerge, bleary-eyed, from the backyard chicken coop. Fuelled with hot coffee, everyone is ready to provide help wherever it is needed. In future years, ladies of the famed "Coffee Brigade"[1] will arrive at the scene of many a Vancouver fire, providing hot coffee and sandwiches to refresh and rejuvenate tired crewmen.

Andy Linton wanders through the debris-strewn streets, hoping to catch any word of the whereabouts of his missing boats. Clambering atop a small mound of debris near the ruins of the

Urn used to serve coffee to fire refugees.

Deighton House, he shades his eyes to a scene of desolation. Everything is black. There is not a clump of green grass to be seen, nor a vegetable patch, nor a solitary yellow buttercup. In the same moment, he feels himself lose his footing. As he struggles to regain his balance, Andy is astonished to see something white and shiny glinting up at him in the morning sunlight. Closer inspection reveals that the something is...*ice*! Beneath the wreckage of the Deighton House ice shed, several giant blocks of ice have survived the fire's intense heat. A thick layer of sawdust encased within the shed's walls has amazingly and effectively insulated the ice as the structure collapsed inward.

With the morning sun rapidly growing hotter, Andy beats a direct path down to his still intact float, grabs up two buckets and hurries back to begin salvaging the ice. In a city where wells have run dry and creeks are ash-choked rivulets, fresh drinking water is as precious as gold.

■ 8 A.M.

As the morning sun climbs higher, one question predominates in the minds of Vancouver citizens. How many lives had been lost? A makeshift morgue has been established in an undamaged office building at the Royal City Planing Mills—a new sawmill under construction near the north end of Westminster Bridge. Inside, twenty-one bundles of remains, some perceived to be human, are laid out on an improvised table of salvaged plywood. Some are little more than charred fragments gathered together in blankets donated from the nearby Bridge Hotel. A note is pinned to each bundle, detailing where the contents were found and under what circumstances.

The office of the Royal City Planing Mills (left) was converted to a makeshift morgue, June 14, 1886.

Grisly remains found at the intersection of Carrall and Cordova streets prove to be those of soda-water merchant Albert Fawcett. Witnesses recollect how a stone-faced Fawcett had driven his horse and wagon at full gallop into the heart of the flames, where no one could have possibly survived. All that is left of the wagon are two tire irons and some melted pieces of soda-water equipment. Mysteriously, the remains of Albert Fawcett's horse are nowhere in sight, leading to speculation that he may have been able to free the animal.

Jackson Abray had fully expected to find the body of his friend and colleague Mr. Anderson, but after carefully retracing the path of their flight down Gore Avenue, he was only able to locate the man's valise—Masonic jewels melted together within.

To his utter amazement and great joy, he later found Anderson alive and well, having somehow managed to crawl to False Creek as the fire abated. After the long hours of being wracked with guilt, Abray's mood is surprisingly upbeat as he wanders the burned-out downtown streets. He comes across Mayor MacLean, making his own assessment of the damage. The two men are on very good terms, Abray having provided financing for Mayor MacLean's run for office during the city's recent election.

With the spectre of looting ever present, Mayor MacLean has ordered Police Chief John Stewart to commission a squad of special constables—able-bodied and respected gentlemen, to provide assistance where needed—despite lack of formal police training. Noticing that a group of men are preparing to paddle away from shore with several barrels of whiskey, the mayor himself quickly turns to Jackson Abray.

"Abray—I'm swearing you in as Special Constable. Your first duty is to retrieve those barrels!"[2]

Mayor MacLean is anxious to prevent any recurrence of the previous evening's rampage, but as a proud native of the Inner Hebrides island of Tyree, one more thing is clear to him: a strong shot of Scotch whisky will do no harm to the morale of Vancouver citizens—in controlled doses.

Harry Devine tramps about the city with camera, tripod and a fresh supply of negative plates carefully stowed in his rucksack. Triumphant that his old photographs of Vancouver have been rescued, Devine has found a new mission—to record the many poignant scenes of a city recovering from disaster. The refugee bivouac at False Creek quickly captures his imagination. A hodgepodge collection of salvaged articles serve a multitude of purposes, as fire refugees make painstaking efforts to re-establish some semblance of home comfort. Clusters of tents have mushroomed overnight—not only the typical, white canvas variety, but ragtag creations pieced together with blankets of every size and colour. Sacks of oats have been put to use as pillows. With a table napkin tucked under his chin, one gentleman is enjoying a meal with a tin plate, knife and fork salvaged from the ashes of Tye's Hardware. He had carefully scrubbed the tableware clean with sand and seawater before putting it to use. Walter Graveley nonchalantly reads a book. Dogs loll on the ground. The bivouac is a remarkable picture of order beginning to emerge from chaos.

Passengers line the deck, aghast at the sight before them, as the sidewheel steamer *Princess Louise* churns into Vancouver harbour on her regularly scheduled run from Victoria. Many of the new arrivals carry trunks and satchels packed to the brim, with settlement on their minds. The captain loses no time in announcing "Free breakfast for victims of the fire!"[3] The ship is scheduled to depart on its return voyage to Victoria at 10 a.m. It will be a crowded vessel, with many individuals quickly abandoning their plans to disembark, and well over two hundred newly homeless Vancouverites gratefully accepting the offer of free passage. Emily Strathie is among them, not overly concerned that she is still clad in her bedraggled print dress and slippers, her normally pinned-up hair askew about her shoulders.

Businessmen Robert McLennan and Edward McFeely eagerly board the *Louise* to seek out their on-board shipment of lumber and construction supplies—stock that had been destined for their half-finished hardware store, now burned to the ground. Misfortune quickly turns to good fortune, as the partners find their wares instantly in demand.

Refugee bivouac near the south end of Westminster Bridge.

■ 11:00 A.M.

Owing to the largely transient population of Vancouver, a precise count of the dead and missing will prove to be a hopeless task. Under the supervision of New Westminster coroner Josias Charles Hughes, a jury of twelve men gathers in the makeshift morgue to view, and with the assistance of witnesses, attempt to identify the bodies gathered thus far. The body of a woman was found on Hastings Street, west of the Burrard Hotel. Another woman's body, with torso still encased in the steel frame of a corset, was found at the corner of Powell and Columbia streets. Four bodies, charred beyond recognition, were found at the corner of Powell Street and Westminster Avenue. The only recognizable body is that of John Craswell. Jumping down a well in a desperate attempt to escape, Craswell had inadvertently suffocated himself. As flames roared over his perceived refuge, they voraciously drew up every last breath of oxygen. James Hartney, who has lost his Ferguson Block grocery, paints an eerie picture in documenting his discovery of Craswell's remains:

> A piece of charred wood was lying across the neck of the deceased and charred embers were in and about the well. The back of the head had the appearance of being burned. There was about one foot of water in the well. Afterwards, in the rear of the Deighton Hotel site, I found the remains of another human body, burnt and charred beyond any hope of recognition.[4]

George Golden, duly sworn in, adds his observations:

> The watch and chain now produced in hand was taken by me to Mrs. Albert J. Fawcett and she stated to me that she recognized it as the watch chain and charm worn by her late husband Albert J. Fawcett.[5]

As no red sealing wax is on hand, the jurors are forced to improvise and affix little pieces of white paper to their official reports with white paper paste.

The search for bodies continues throughout the city. While sifting through the ashes of the McCartney brothers' drug store, one group of searchers uncover a blackened skeleton.

"Poor fellow!" one of them innocently remarks. "He must have been sick before he died! His back is all wired together!"[6]

Unknown to everyone present, the "victim" is actually "Jimmy," the remains of a Swedish man who had hung himself back of Moodyville two years previously. Desiring to set up a display skeleton for the small school of anatomy that he had recently established in his office and surgery above the drug store, Dr. Henri Langis had rowed over to a small graveyard on Deadman's

[CORONER.]

CORONER'S PRECEPT.

New Westminster District
Vancouver City 13 } To the Constable of Vancouver City

By virtue of my office, these are, in Her Majesty's name, to require and command you, immediately upon sight hereof, to summon and warn ~~twenty-four~~ Twelve good and lawful ~~men of~~ to be and appear before me, Jonas Charles Hughes, one of the Coroners of New Westminster District at Vancouver in the said District on the 14th day of June, at 11.30 of the clock in the fore noon, then and there to enquire of, do, and execute all such things as on Her Majesty's behalf shall be lawfully given them in charge, touching the death of John Craswell & others AND be you then and there to certify what you shall have done in the premises; and further to do and execute what in behalf of our said Lady the Queen shall be then and there enjoined you.

GIVEN under my hand and seal, the 14th day of June, A. D. 1886

J. C. Hughes
Coroner

Coroner's report, June 14, 1886.

Island and quietly exhumed the remains. As Dr. Langis is away on business in New Orleans and unavailable to identify his property, "Jimmy," along with other unfortunate souls having no known family or friends, will be solemnly buried in an unmarked grave.

Walter Graveley aimlessly wanders about the ruins, like many others, searching for anything salvageable. Before long, a stranger approaches him and says, "I have some papers belonging to you!" Astonished, Graveley receives an ash-smeared bundle, which, upon closer inspection, proves to be a collection of deeds and land agreements with the unmistakable letterhead of Graveley and Spinks. Amazingly, everything is intact, the typed and handwritten documents still clearly legible.

"Where did you get these?" Graveley asks.

"I found them on the beach. They were inside a tailcoat."[7]

It seems utterly impossible to Graveley, but indeed true that his tailcoat, along with all of its contents, has somehow managed to survive the fiery collapse of the Sunnyside Hotel. Had some looter gone from room to room as the fire bore down, grabbing articles at random? Was the tailcoat blown out of the window as panes of glass shattered with the intense heat? He can only theorize. As he continues with his scavenging, he manages to uncover two reasonably intact dry goods boxes. Not far from the ruins of Graveley and Spinks, he positions the boxes to form a makeshift desk, sits down with his rescued documents, and begins to draw up leases for new businesses along Cordova Street. Before long, the ever-enterprising Walter Graveley has a steady stream of interested customers.

Tales of strange and unlikely cases of survival are told and retold all over the city. At the southeast corner of Abbott and

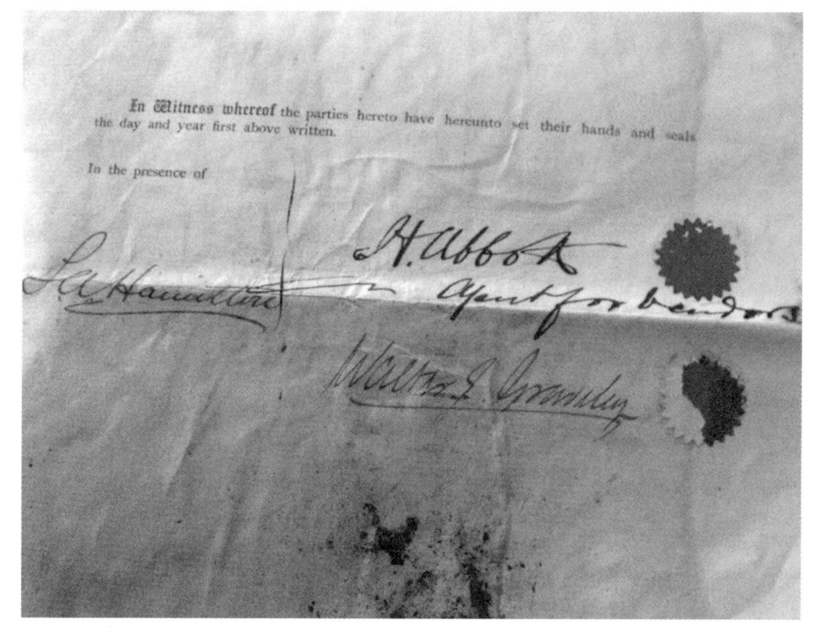
In Witness whereof the parties hereto have hereunto set their hands and seals the day and year first above written.

In the presence of

L. A. Hamilton

H. Abbott
Agent for Vendors

Walter E. Graveley

Ash-smeared document belonging to Walter Graveley.

Hastings streets, a small shack stands perfectly intact amidst a wilderness of destruction. Rumour has it that the sick, elderly man who lived there, fought the fire as hard as he could. Eventually overcome by heat and smoke, he crawled inside his home and closed the door, fully expecting to die—only to re-emerge several hours later, mystified, but alive and unscathed.

Two horses belonging to Hiram McCraney have survived, thanks to the kindhearted efforts of a person or persons unknown. The animals had been led to a deep cutting in Cordova Street near the future intersection of Granville, and tied to wagon wheels. Of two cows belonging to Reverend Hall, one has died, while the other wanders aimlessly along the Burrard Inlet shore, vainly searching for clumps of unburned grass.

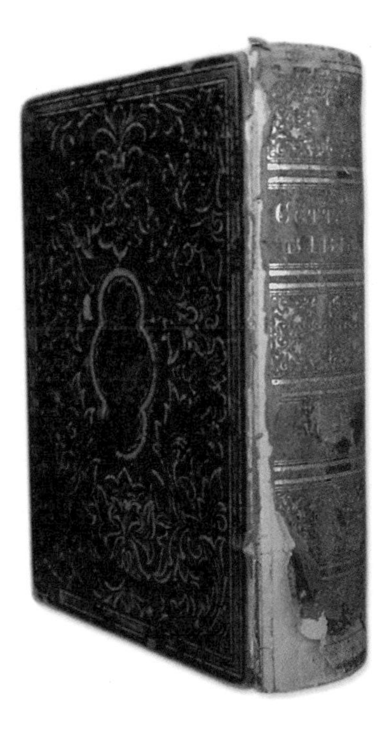

The Macey family Bible.

The Maceys draw up their undamaged family Bibles from a well. Christina Reid is delighted to find her cooking pot and sewing machine in near-pristine condition amidst the ruins of her house. She muses that the thick coating of varnish on the machine's cabinet somehow managed to repel flying cinders. In days ahead, the Reid sewing machine will be circulated from one family to the next, stitching everything from patches on burned clothing to blanket tents.

Cast-iron cook stoves throughout the city have taken damage amidst the inferno, but are potentially salvageable. Tin chimney pipes are beyond repair, as are built-in fixtures of glass and porcelain, but stovetop burners and cast iron fry pans are carefully

Reid family cooking pot.

gathered up for makeshift use. Kitchen wood boxes with their ever-ready supply of kindling and tinder are long gone. The Sunnyside Hotel safe has been located twelve feet under water, its contents thoroughly drenched, but otherwise intact.

At Hastings Mill Store, Calvert Simson hurries up and down the stairs, his arms laden with wooden boxes of canned meat, bread, cheeses and sea biscuits. Everything will be offered up, free of charge, to any and all in need. With the steady demand, shelves are rapidly emptying of all food stocks, save for the vast supply of sea biscuits. Appetites are returning full force as hard work and determination replace the initial shock of the fire. George Keefer, a CPR man who had been supervising a large crew of workers on the rail bed between Port Moody and Vancouver, makes arrangements for his company scow to be towed to the mill dock. The scow is a treasure trove of supplies—food, fresh water, tents, flooring, tables, benches and a wealth of other necessities. Before long, the crew erects two large tents and strings a canvas banner overhead, reading "R.R. Dining Rooms." While not wishing to profit from the misfortune of others, Keefer hopes to recover a portion of costs while offering "First class meals to all for the price of 25 cents."[8] The tents are soon filled to capacity with grateful diners.

An incredible scene presents itself at the intersection of Carrall and Cordova streets. The heart of Vancouver's business district has been reduced to two white canvas tents, surrounded by a wilderness of black ash. Harry Devine positions his tripod to capture the tents, with the irony of an unscathed Regina Hotel off in the distance. Historical significance is also on the mind of James Ross. Quick inspection had confirmed that his *Vancouver Daily News* office and all of its contents were destroyed in the inferno, the printing press melted beyond recognition. Ross is travelling to Victoria—busily recording the stories of fellow survivors aboard the first vessel on which he has been able to secure passage. Upon arrival in the capital city, he will head straight to the J.B. Ferguson Book and Stationery Store, in hopes of purchasing a small hand printing press.

Calvert Simson

While some individuals endeavour to find a silver lining amidst the ashes, others in the city dismally tally their losses. Frank

Two white canvas tents, surrounded by a wilderness of black ash, June 14, 1886.

Hart's furniture store is a complete write-off, in terms of both structure and inventory. Thomas Dunn estimates that he has lost $2,000 in stock, along with his home and a new piano just arrived from Victoria. Edward Gold chastises himself repeatedly, while picking through the ruins of his Water Street general store. Only three weeks prior to the fire, insurance policies amounting to $10,000 on his stock had expired. No insurance company had been willing to renew, based on the potential for fire damage because of the close proximity of burning slash. In recent days, Gold had begun formulating plans to relocate half of his stock to New Westminster and open a new store. Now, in addition to losing his business, he is penniless.

Father Clinton, fully occupied with providing comfort to stricken families, has little time for sadness or contemplation. In the priest's absence, lumberman Donald Charleson, who also serves as the warden of St. James Church, picks through the sanctuary ruins. Searching through heaps of burned timber, he uncovers the blackened remains of an altar chalice. The chalice is part of a set of altar vessels given to the Vancouver congregation from the parishioners of St. James Church in the English West Midlands township of Wednesbury.[9] Continuing his search, Charleson soon uncovers a misshapen, metallic object, which appears to be dripping with hardened globules. It is not long before the awestruck warden realizes that he is holding the melted remains of the St. James church bell—that same bell which Father Clinton had tolled so heroically only hours before. However ugly and useless the bell now is, Charleson quickly recognizes that its historical value will be significant, and inwardly vows to ensure its care.

Melted remains of the St. James Church bell.

Several others come across items amidst the destruction that are hopelessly beyond repair but of immeasurable sentimental value. Poking amongst the ashes of her husband's downtown office, Jessie Ross uncovers some mangled cutlery and the charred remains of her wedding bouquet holder. Anna McNeil finds several pieces of jewellery—once elegant golden earrings and brooches, now melted into shapeless blobs. Years later, Margaret Florence McNeil will donate her mother's treasures carefully wrapped in a yellow handkerchief to Vancouver City Archives. Hiram Scurry has lost his barber shop, along with all of its contents—save a shaving razor, which he gratefully pockets.

■ 6 P.M.

The steamer *Mastick* departs from Seattle on an unscheduled charter voyage to Vancouver. The *Seattle Times*, following up on information circulated by the mother of Sherman Heck, has issued a press release:

> Vancouver in Ashes. The "City of Big Hopes" Swept by Flames. Among the Seattle parties who are interested in Vancouver and will suffer more or less from the fire are: P. Frederick, Clayton and Silvain, T.B. Mangan, Sutherland and Co., Sherman Heck, Mr. Calson, A.E. Wayne, Chas. Norager and J.H. Carlisle.[10]

Over the day, grey clouds have steadily been gathering. As the first drops of rain in over three weeks begin to fall on the wreckage of Vancouver, Mayor MacLean and Thomas Dunn hunt for a place to take shelter. Mayor MacLean is emotionally and physically exhausted. He has personally lost two business establishments, a quantity of fine furniture which had recently been

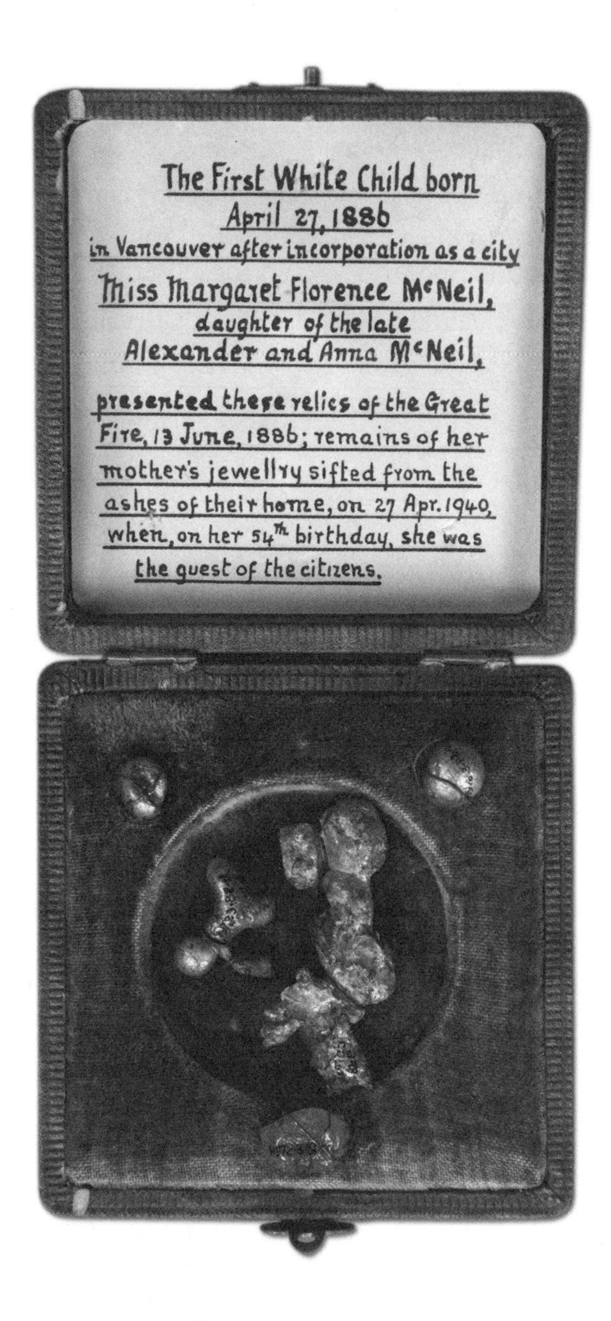

Anna McNeil's melted jewelry.

Melted cutlery found by Jessie Ross.

shipped from San Francisco, and he has not slept in over twenty-four hours. The men make their way to the Regina Hotel but find it full to capacity. Unwilling to use their social status to trump the needs of others, they backtrack to the *Princess Louise*, once again moored at Hastings Mill. Staterooms are full, and virtually every spare bit of space is taken up with tired refugees. Two large tables in the ship's dining room have thus far been unclaimed. Chuckling for the first time in a long while, the mayor and his colleague engage in a friendly game of tug-of-war to see who will win the solitary table cloth for use as a bed-spread.

In the Days Ahead

White canvas tents have sprouted throughout the city as Vancouver property owners dig about the ashes to find the iron survey stakes identifying the exact locations of their lots and re-claim their land. In a tent erected at the north foot of Carrall Street, a grim-faced Mayor MacLean calls his aldermen to order. Everyone is well aware of the mayor's eerily prophetic declaration, now all but taunting them from Vancouver's first city council meeting minutes:

> We require immediately protection from fire, and any delay on this matter endangers a large amount of valuable property.[1]

Alderman Lauchlan Hamilton has jokingly lightened proceedings by attaching a hand-painted "City Hall" sign to the tent's ridgepole. Sequestering themselves within their rudimentary shelter, Vancouver's leaders attempt to carry on with city business in as orderly a fashion as possible. Atop everyone's list of priorities is a formal inquiry into the cause of the fire and full extent of the damage. It is clear that many prominent Vancouver businessmen have suffered appalling losses. Anywhere from six hundred to one thousand buildings have been totally destroyed. Few business owners have insurance. Those that do hold policies are discovering that only a fraction of their property values will be compensated. The city will require huge infusions of cash to rebuild.

News of Vancouver's calamity soon spreads farther afield. The simple, yet dramatic wording of Mayor MacLean's telegram captures the headline-hunting attention of newspaper editors from several major Canadian dailies. "Vancouver in Ashes" appears on the June 15 front pages of the *Globe*, the *Montreal Gazette* and the *Manitoba Free Press*. The June 15 *Toronto World* includes a personal plea from Mayor William Howland, an ardent philanthropist:

> Mayor Howland requests all who may have anything to subscribe to let him have it as soon as possible, as he desires, in the name of Toronto citizens, to telegraph some money today. A dollar today may be worth ten a week hence.[2]

Relayed via Philadelphia, the news will soon reach New York and London, England. New Westminster's June 16 issue of the *Mainland Guardian* publishes a graphic account of the fire, compiled by an unnamed "Special Correspondent." Arthur Herring, having safely returned to New Westminster with his family, relates a compelling eye-witness description:

> The people had time enough to escape, but they hesitated and delayed, hoping to save trifles. I heard the crackling of the great

> fire coming through the woods, and I returned to False Creek. On the way back, I saw a child at the window of a cabin. I stopped and gave the alarm. I said to the mother, "Come quick, the fire is coming this way, you have no time to spare," and I took the child by the hand. The woman said "Wait until I put on his boots." I told her there was no time; the fire was coming; I took the child by the hand and ran; the woman waited; I looked back, and the smoke had gathered around the cabin. I did not see her any more. I left her child with a lot of women who were gathered at False Creek bridge, they knew the woman. I do not know whether she escaped or not.[3]

When Mayor MacLean's telegram to Sir John A. Macdonald remained unanswered, on June 15, a second, more curtly worded telegram is sent by Noah Shakespeare, a federal MP for Victoria and distant relative of William Shakespeare himself:

> Vancouver City reduced to ashes, three thousand people houseless and destitute provincial Govt & corporations assisting what will your government do please answer.[4]

The Collins Overland Telegraph[5] line had been Vancouver's link with the outside world for many years. While much of the line had been abandoned by 1886, a single wire, strung tree-to-tree, follows the line of least resistance to New Westminster, beyond to the Sumas border crossing and points eastward. While the line had proved to be an efficient means of communication between Vancouver and Fraser Valley residents in times past, response is agonizingly slow from Ottawa. In exasperation, Mayor MacLean forwards yet another telegram on June 22:

> Our citizens are in extreme want will your Gov't please do something for us.

This communication finally generates the desired result. A reply arrives from Prime Minister Sir John A. Macdonald:

> M.A. MacLean, Vancouver City, B.C. Dominion Government will contribute five thousand dollars.

Mayor MacLean telegrams back:

> Message received please accept our unbounded thanks for your most generous and timely assistance.[6]

A telegram arrives for CPR Superintendent Henry Abbott:

> Relieve any distress in consequence of the fire. You are authorized to contribute on behalf of the trustees of the town-site an amount not exceeding $3,000. W.C. Van Horne.[7]

Help begins to arrive—as a trickle at first, soon a flood tide. More wagons rumble along the New Road from New Westminster, loaded with supplies and cash donations. Boats arrive in a steady stream from Port Moody, Victoria, Nanaimo and points south of the border.

The Government of British Columbia donates $1,000, as does the city of Toronto. Hamilton, Ontario and Victoria, B.C. donate $500 respectively. Despite their losses, many Vancouver citizens offer their own contributions. Henry Abbott personally pledges $500, grocer David Oppenheimer, $100. In addition to free lumber, Richard Alexander offers free meals in the Hastings Mill cookhouse. George Keefer donates one ton of flour.

A mind-boggling amount of items has been lost—not only critical rebuilding supplies, but all manner of everyday household goods. Furniture, kitchen crockery, utensils, tinware, lanterns, bedding and baby diapers are high on many a wish list.

Books, toys and other more simple items of comfort—though not as urgent a priority, are not to be forgotten. Traumatized Vancouver children receive a welcome diversion with the arrival of several buckets of homemade ice cream by steamer from Nanaimo. Schoolwork has been set aside. Although the Hastings Mill schoolhouse has survived the fire, families have scattered and an early start to summer vacation is affirmed with few protestations.

Richard Alexander

Perhaps nowhere has Vancouver's plight been more deeply felt than by the citizens of Moodyville. The north shore community's sympathy, thoughtfulness and generosity so move Mayor MacLean that he is compelled to write a letter of thanks to the local schoolteacher, Margaret Thain:

> Dear Madam: As Mayor of the City of Vancouver, I wish on behalf of its inhabitants to tender yourself and the kind ladies of Moodyville, our sincere thanks for your liberal contributions of money and clothing in aid of the sufferers by the late fire here. Also allow me to express my sense of our deep obligation for the generous hospitality extended by the people of your village to our suffering citizens on the evening of Sunday last, and even up to the present time. The assistance thus afforded, will, I assure you, be not soon forgotten, but its memory will always remain with us, side by side with the remembrance of our great calamity. Again, thanking you and those ladies who have so kindly assisted you. I am, dear Madam, Yours very gratefully, M.A. MacLean, Mayor.[8]

He encloses a second letter for Mrs. Thain's daughter:

> Dear Miss Thain: Your very kind gift of pillows from "The Little Helpers" was received by me and distributed according to your instructions. In addition to the other aid offered to our destitute citizens by the inhabitants of your village, in the way of shelter, food and money, your contribution for their benefit proves you to be worthy children of worthy parents. The assistance thus given to us in our hour of need will not soon be forgotten, I assure you.[9]

While in Victoria to shop for a new printing press, Ross met up with his good friend and journalist colleague, the Honourable John Robson, provincial MLA for the district of New Westminster. Robson is a former editor of New Westminster's the *British*

Moodyville, c. 1885.

Columbian, and offered to write a letter of reference that would allow Ross to run off copies of the *Daily News* on the *British Columbian* press. In Robson's estimation, this arrangement would give Ross plenty of time to construct a sturdy new printing plant and replace other much-needed supplies. Realizing that the interior of a leaky canvas tent pitched amidst heaps of ash would not be an ideal location for running off quantities of newspapers, Ross gratefully accepted the idea.

Having returned to Vancouver, Ross finds a secure location to store away his newly-purchased printing press for future use. Armed with a supply of new pencils and fresh writing paper, he then begins to scour the burned-out streets for eye-witness reports. It is not difficult to find fire victims ready to talk about their experiences. With slipper-clad feet propped in easy chairs, Charles Gardner Johnston and John Boultbee quietly describe their harrowing escape. Individuals who sought safety aboard the *Robert Kerr* indignantly relate how "the watchman on the vessel, with all the proverbial insolence and stupidity of insect authority,"[10] attempted to prevent them from boarding the ship. Frank Hart announces that he will be re-building his furniture store on its old site. Before long, Ross has accumulated a wealth of graphic recollections and a full-page list of businesses in the process of re-construction.

Certain Vancouverites do not have a way with words. Lauchlan Hamilton's survey partner John Leask, penned a simple June 13 entry in his 1886 Survey Diary: "City of Vancouver totally destroyed by fire." Newspaper editor James Ross is on a different wavelength. On the evening of June 16, Ross rides a borrowed horse bareback to New Westminster and works through the night, writing up the full account of the fire to appear in a Vancouver-based newspaper. With the assistance of The *British*

Columbian's Sid Peake and typesetter Robert Matheson, Volume One, Issue Twelve of the *Daily News*, a double-sided, eight-inch-by-ten single sheet of print, is readied for delivery to Vancouver. On Thursday, June 17, there is a brief lull in the pounding of hammers and drone of saw blades, as issues of the *Daily News*, fresh from New Westminster, are circulated amongst the townsfolk. Details of the events of June 13, simply headlined "THE FIRE," have been captivatingly recorded by James Ross:

> Probably never since the days of Pompeii and Herculaneum, was a town WIPED OUT OF EXISTENCE so completely and suddenly as was Vancouver on Sunday. All the morning the usual pleasant breeze from the ocean was spoiled by smoke from fires in the portion of the townsite owned by the C.P.R. Co, west of the part of town already built, but no alarm was felt in consequence. The place wherein these fires existed was until two or three months ago, covered with forest. A large force of men had been engaged in clearing it. The trees were all falled, and the fallen trees, stumps, etc., were being disposed of by burning here and there in separate heaps. A few weeks ago, during a gale from the west, the city was filled with smoke and cinders from these fires, and fire reached close to several outlying buildings, but after some fighting danger was averted. This, doubtless, tended to lull people into a sense of security on Sunday. It was about two o'clock in the afternoon that the breeze, which had been blowing from the west BECAME A GALE.
>
> Persons living near the Harbour and in the eastern part of the city hurried toward the wharves at the Hastings Mill, and crowded upon the steamers moored to the wharves. On the steamers and wharves, while the city was a mass of roaring flames, where gathered hundreds of frightened and excited men and sobbing women and children. Anon, there emerged from the dense smoke one and another, GASPING AND BLINDED, with singed hair and blistered hands and faces, who had strug-

> gled almost too long to save property. A considerable number of people were surrounded by the fire and cornered near J.M. Clute and Co's store, and their only means of escape was to make rafts of the planking in a wharf at that place, and push out into the harbor. The wind was blowing fiercely, making the water rough, and the party were in no little PERIL OF DROWNING. They made their way to a vessel which was at anchor in the harbor, and the watchman on the vessel, with all the proverbial insolence and stupidity of "insect authority", refused to let the party come aboard. He was very soon perceived, however, that his refusal "did not count" and that his life would "not count" for much if he attempted to keep people off the vessel, and surrendered unconditionally.

James Ross makes a solemn promise to his readership:

> Like nearly all others who had started business in the new city, however, we perceive that the fire, whatever may be its effect upon individuals, is to the city as a whole not a very serious matter, in fact it can scarcely impede the progress of Vancouver at all. A few months, or even a few weeks, will restore the city to as good a basis as it was on before the fire. We have therefore determined to continue the publication of the *Daily News*. It will appear in reduced form (we hope, however, to present four pages in a few days) until new material can be obtained.[11]

It has been a daunting task to produce the newspaper a mere three days after the fire, but the results are gratifying for James Ross. The *Daily News* becomes a morale booster, synonymous with the city's recovery. A second issue, published on June 18, describes a city rapidly beginning to heal its wounds. Jonathan Miller has set up a temporary post office at the Royal City Planing Mills and has not missed a single delivery of mail, despite having to hunt down various addressees. Seth Tilley has re-established his telephone office at the Bridge Hotel. Names of individuals on the newly formed Relief Committee are published, along with a list of financial donors. In an article boldly emblazoned "CORRECTION," Captain Dyer strongly refutes the claims that he had attempted to bar people from climbing aboard the *Robert Kerr* during the height of the crisis, and had "simply ordered back some, who in their excitement crowded up the ladder too many at a time, and were in danger of crowding each other into the water."[12] Thomas McGuigan is given high praise for saving the City Archives:

> A good deal of embarrassment would have resulted to the authorities if the documents had been destroyed, and Mr. McGuigan deserves credit for his thoughtfulness.[13]

Minutes

1 The First Meeting of the Board of
Aldermen of the City of Vancouver was
held in the Court Room, Vancouver on
Monday May 10th 1886 at 2 o'clock P.M.
2 His Worship the Mayor Mr M.A. MacLean
presiding. Present Aldermen R. Balfour
C.A. Caldwell, P. Cordiner, Wm Dunn
J. Griffith, J. Humphries, H. Hemlow
E.P. Hamilton, L.A. Hamilton and
J. Northcott.
The above named Aldermen were duly
3 Sworn in by His Worship the Mayor and
their declarations of Office filed by the
Clerk.

First Vancouver City Council meeting minutes, saved by Thomas McGuigan.

Harry Devine is commended for his rescue of the photo negatives. There is a poignant tribute to the memory of the immensely popular soda-water man, Albert Fawcett:

> His remains were found near his wagon, and he probably lost his life attempting to save his horse, which was unhitched and ran down to the harbour and into the water. The animal was badly burned but will recover.[14]

Heroes are praised for their efforts on behalf of the stricken city, while others are roundly admonished:

> A few despicable characters attempted to do some looting during the fire, but it's probable they didn't get away with much.[15]

"Raised from the Ashes in Three Days!"[16] proudly proclaims a cotton banner over Duncan McPherson's rebuilt, if somewhat rustic, CPR Hotel. The Tremont Hotel is soon to follow, albeit appearing more like a logger's shack than a proper rooming house. For the time being, various proprietors are more concerned with having a shelter for their guests to drink alcohol than a multi-room structure with feather beds and carpeted floors. Via the *Daily News*, socially conscious Captain William Clements tries to dispel rumours of anything untoward happening within his establishment:

> Captain Clement of the Tremont Hotel is dispensing, among other beverages, milk of the best quality.[17]

The policing of belligerent drunkards clearly presents an ongoing challenge. Jackson Abray and his special constable colleagues wear makeshift badges fashioned from American silver dollars, roughed down on one side and engraved "Vancouver City Police," as they carry out their duties. With no adequate

The Tremont Hotel, open for "business" after the fire.

prison facility close at hand and little mood amongst hardworking Vancouverites for tolerance, troublemakers are unceremoniously handcuffed to a logging chain fastened around a newly installed telegraph pole outside the tent city hall. Here they while away the hours, chewing tobacco and, for amusement, spitting the juice at flies landing on a nearby public notice board. If flies are lacking, a small slit in the tent becomes the target. Before long, Vancouver's temporary headquarters for official city business has become awash with tobacco juice stains.

Reconstruction continues at fever pitch. One unforeseen advantage of the fire is that it has carved through much of the mountains of debris on the outskirts of town in one fell swoop,

Jackson Abray's silver-dollar Vancouver City Police badge.

effectively reducing weeks of hacking, sawing and controlled burning. Crewmen descend on the fire-ravaged areas with newly-shipped grub hoes and shovels, pulverizing the charcoal remains of long dead tree branches, distributing the ashes over a wide area and covering them with soil.

Everyone is saddened to learn that the Princess Louise Tree,[18] a landmark Douglas fir at the north foot of Gore Street, is badly scorched and cannot be saved. When the Marquis of Lorne and his wife, the Princess Louise—fourth daughter of Queen Victoria—visited Hastings Sawmill in 1882, plans had been made to cut the tree down for their entertainment. The princess, a passionate artist and sculptor, was captivated by its beauty and asked if it could be spared. Her wish was granted, and the Princess Louise Tree, so named in her honor, had remained—majestically tall and undisturbed amidst the rapidly depleting stand of timbers flanking the mill.

The delicate question of allowing purchased lots to be cleared by fire is put before city council. After lengthy discussion, no formal resolution is reached, but everyone is in general agreement that the risks outweigh the advantages. No further burning will be permitted until the city's new fire engine arrives from Ontario and three large water tanks have been constructed. Fresh planking is laid down along Cordova Street to provide a firm foundation for the ever-increasing numbers of wagons jolting between construction projects.

A Relief Committee is formed, comprising the mayor and aldermen, Frederick Innes, John Blake, Richard Alexander, Alfred Ferguson and John Rankin. As chairman of the committee, Mayor MacLean has the daunting task to assess which Vancouver citizens are in direst need. For many fire victims, a hammer, a saw, a ladder or even a solitary nail are as rare as the cash to purchase them. Those devoid of funds are directed to apply to the mayor himself with an enumerated list of their losses. Some try to capitalize on the offer of free supplies, a tactic not unnoticed by the astute MacLean. One applicant, a self-proclaimed carpenter who has "lost everything," wants a quantity of tools to enable him to rebuild a house. Pressed for specifics, he gets as far as "a hammer, saw and chisel," before becoming hopelessly muddled in his efforts to name some of the more common carpenter's tools.

"Are you sure you had all these tools you mention before the fire?" the mayor asks suspiciously.

"Oh yes, your honour, and more than that!"

"Then take this," Mayor MacLean says with a sigh, handing him an order for tools from Thomas Dunn's newly shipped hardware supplies, "and may the Lord have mercy upon your soul!"[19]

Vancouver's first mayor is a common sight in the city from

day to day. Malcolm MacLean's May 3 election victory by seventeen votes, in an often bitter campaign with Hastings Mill manager Richard Alexander, was questionable at best—but seemingly overnight, the mayor has become a stalwart leader, well-respected by everyone.

On the afternoon of Sunday, June 20, exactly one week after the fire, Reverend Thompson stops by one of the Cordova Street crews. Politely, he asks if they would "consider halting their work for a few minutes and give thanks to the Almighty for His divine grace in sparing their lives the previous Sunday."[20] Perhaps more out of guilt than desire, the men lay down their tools.

Mayor Malcolm MacLean

Within the partially-completed reconstruction of George Allen's shoe store, makeshift pews are fashioned from planks laid across empty spike kegs. As the reverend begins to speak, Mayor MacLean tiptoes in and quietly takes a seat. The service is short and simple, with no singing of hymns or taking of collection. It is purely an expression of gratitude for lives spared and opportunities renewed. As it ends, the mayor rises and shakes hands with each and every member of the congregation.

Another constant presence on Vancouver streets is Emma Alexander, president of the Women's Relief Committee. While Emma's home and worldly possessions have survived, most of her good friends and neighbours have lost everything—a fact that weighs heavily on her mind. The Alexander cook stove is put into near-constant operation, turning out quantities of fresh home baking and savory stews. Every spare item of bed linen is converted to much-needed bandaging and soothing compresses. Emma pays little heed to the lack of city boardwalks and the quagmire of mud and manure underfoot as, with basket in hand, she gently calls through one tent flap after another, inquiring on the wellbeing of the inhabitants within.

The *Robert Kerr*, securely anchored offshore in the harbour, continues to shelter Vancouver citizens with no place to call home. Captain Soule, his wife and children, having moved aboard, are happy to consider their vessel "common property"[21]—a place where individuals can rest weary limbs after a day of back-breaking labour. Meetings to discuss relief efforts are held within its forward cabin and fortifying meals are served up from the galley. Theresa Soule becomes an adept rower and proudly enjoys her new nickname, "The Muscular Christian,"[22] as she makes her frequent journeys between ship and shore.

Although most Vancouverites have little in the way of possessions, various "lost and found" articles are turned in to the Relief Committee. The June 24 issue of the *Daily News* reports unclaimed goods including "2 tin-covered and 1 leather-covered trunk, 1 small carpenter's chest, 2 leather valises and 1 bundle of clothing are at the Relief Committee tent waiting for their owners."[23] To the delight of many, Harry Devine has quickly re-established a makeshift photo studio. His simply worded advertisement in the *Daily News* quickly results in photos developed from the rescued negative plates becoming much-sought-after treasures:

> Photographs Vancouver Before and After the Fire. Photographs of the City of Vancouver before and after the fire can be had at our tent on Cordova Street. Parties desiring the same will please give us a call.[24]

Andy Linton, having managed to re-group his rental boats, allows them to be freely used for journeys back and forth across Burrard Inlet as needed. Drinking water is in high demand under the hot summer sun, much of it being supplied by boat from Moodyville. Concerned about his well running dry, John Cartwright, proprietor of the Bridge Hotel, decides to chain up the pump—a decision that is rapidly reversed when angry and thirsty Vancouverites descend upon his premises. Many Vancouver wells are still ash-clogged, and good hygiene standards are not being prioritized as they should be. Much to the dismay of Vancouver doctors still attending to the needs of burn victims, the first cases of typhoid are beginning to emerge.

James Ross continues to scour the city, searching for stories both inspiring and disillusioning. Whenever he comes across any-

Harry Devine's camera, c. 1886.

thing that goes against the grain of Vancouver's recovery, he is quick to publicly vent his ire in scathing editorials:

> The Relief tent is daily besieged by individuals the most ungracious and uncivil and the patience of the Mayor is often sorely tried. Some people were on the principle "the more I get, the more I want," while a great many of the more deserving, on account of their modesty in asking aid, are not receiving their due.[25]

On June 29, one of James Ross's competitors, the *Vancouver Daily Advertiser*, resumes publication from a tent on Carrall Street—the first Vancouver newspaper to be published within the city, post-fire. A front page proclamation reads,

> Our immediate prospects are indestructible. We will rise again, superior to all difficulties. *Fortis in arduis*.[26]

Ross's hopes of continuing with daily publication of his newspaper are not to be realized in the short term. The arduous schedule of collecting stories by day, commuting to New Westminster and churning out copies of the *Daily News* by night quickly takes its toll. His business partner, R. Harkness, has decided to resign. After managing to produce issues sporadically through to July 2, Ross is temporarily forced to suspend publication.

Two weeks after the fire, work crews fanning out from the reconstruction areas discover a partially burned mattress with a badly decomposed body lying beneath. There is little doubt that it is yet another fire victim—but with no means to identify the body and construction progress beckoning, the decision is made to bury the body and mattress exactly where they have been found—beneath the charred earth of Hastings Street, a few yards west of Carrall.

Despite many a setback, Vancouverites continue to plod doggedly forward, slowly reclaiming their city. The prevalent mood can perhaps best be summed up in Mayor MacLean's letter to a friend in Montreal, later reprinted in the *Vancouver News*:

> Come along here. It will do you good. We are wiped out by the fire, but in the space of two weeks two hundred houses are underway. Our people have energy enough for anything; they are a live people, they have a live mayor, they work together for the common good, and to each one in particular. We have had a terrible clean out but before the end of the year we will have 10,000 people here.[27]

In the Months that Follow

History-making events continue to unfold in and around Vancouver. The July 4, 1886 arrival of the first through Canadian Pacific Railway train to Pacific Ocean headwaters at Port Moody draws a large crowd to Burrard Inlet's eastern extremity. One hundred and fifty passengers disembark, many with aspirations to continue on to Vancouver via boat or carriage. The city is more accessible than ever before. By July 23, James Ross has a new printing plant up and running in a newly constructed, one-storey building on Cordova Street. In addition to publishing the newly renamed *Vancouver News*, James offers general print services:

> We have the best facilities for doing every description of plain and ornamental printing. Coloured work a specialty.[1]

On July 24, the city's first fire by-law appears on the front page of the *Vancouver News*:

> THE FIRE BY-LAW. A Summary of Its Clauses—The City Fully Protected Under Its Provisions.[2]

The new rules are strict and uncompromising:

> Previous to the erection of any building, it shall be the duty of the person about to build the same, to notify one of the Fire Wardens, who shall immediately inspect the ground in order to ascertain if the provisions of the Fire By-Law have been complied with.
>
> All rotten wood and decayed vegetation are to be removed promptly from the piece of land upon which a building is to be erected.
>
> No fire shall be kindled in any street, alley or vacant place within the city.
>
> Fire or live coals shall not be carried through the streets, unless placed in a metal pan or vessel.
>
> Lighted candles or lamps shall not be used in any stable or place where combustible materials shall be stored, unless well secured in a lantern.
>
> Lighted cigars or pipes shall not be used in any building where combustible material may be.
>
> Straw, hay, etc. shall not be put in a stack or pile, without being securely enclosed so as to protect them from flying sparks.
>
> No more than 10 pounds of gunpowder shall be stored in any house at any one time, unless the same is kept in a magazine.
>
> Every chimney or flue shall be built of brick or stone four inches in thickness and must be cleaned once every six months.

> Chimneys, flues, etc., built in contravention of this by-law may be pulled down or removed at the expense of the owners.
>
> A volunteer fire brigade of not more than 50 men, shall be organized, who shall be under the control of the Chief, who shall be elected from amongst them annually.
>
> Any fireman injured while engaged on duty shall be compensated, and the widow (or orphans if any) shall be entitled to such pecuniary aid as the Council may determine.
>
> The owner or occupant of every building shall supply a ladder of sufficient length to reach the roof.
>
> Buildings may be pulled down or demolished whenever necessary to prevent the spreading of fire.
>
> It shall be the duty of the Fire Brigade to perform such services as may be required of them by the chief officer in charge, and not to depart from duty without permission.
>
> In case of fire or danger of fire, it shall be the duty of every able-bodied male inhabitant to assist to the utmost of his power in preventing or suppressing the fire.

In total, Vancouverites are legally bound to comply with thirty-eight new stipulations. The list ends with an ominous warning:

> Any person found guilty of not heeding to any clauses in this by-law, may be fined up to $100 or committed to prison for up to six months.

The Chief of Police and all city constables are granted special authority as Fire Wardens to ensure that the new regulations are strictly obeyed. Any home or business can be entered at any time for inspection.

Vancouver city police force, 1886. Left to right: Const. Jackson Abray, Chief Const. John Stewart, Const. V.W. Haywood, Sgt. John McLaren.

The Relief Fund has grown sizably, and with it, the controversies regarding its dispersal. A letter to the editor in the same July 24 *Vancouver News* issue containing the Fire By-Laws, reads:

> I understand that about $20,900 has been received so far by the Mayor and Relief Committee in aid of the sufferers by the fire, and as I also understand that your columns are open to the pub-

lic to discuss matters of importance, I ask you to call for a full statement of monies received and expended; the names of recipients and the amount received by them. In making this appeal to you I am merely expressing the wishes of our citizens in general. Let the accounts be audited and published as soon as possible not only for our own behalf but for the benefit of our friends in the east who have come forth so nobly to help us. I am sure that our worthy mayor will take the hint and see the advisability of this proceeding. Trusting you will find room for this in your enterprising journal. I am, Mr. Editor, Yours very sincerely, PARS PRO TOTO.

A somewhat terse reply is printed beneath the letter:

> We have reserved all comments upon this matter until the auditors Mesers. G H. Ham and John Boultbee, who were appointed by the Relief Committee on Monday last, had completed their examination. We trust that those gentlemen will get to work as quickly as possible and present the report which the citizens of Vancouver and the donors to the fund scattered all over the continent have been looking for so long.[3]

New, occasionally unlikely opportunities for business continue to present themselves in Vancouver, often as a direct result of the fire. Although much paper money and coinage was abandoned and destroyed on June 13, a few cast iron safes have proved remarkably sound. Mr. G. Robertson publishes a glowing testimonial to the durability of Goldie and McCulloch safes:

> FIRE! FIRE! TESTIMONIAL. SIR—We, the undersigned, having witnessed the terrible fire which wiped out of existence the young city of Vancouver on the 13th inst., have after the fire, examined the safes sold by you, some of which were put to the most severe test, having been surrounded by large quantities of lard and bacon, and we are pleased to testify to the remarkable

> manner in which every safe sold by you preserved its contents, not only books and papers, but also thousands of dollars in paper money were taken out in perfect condition. We can therefore, with the greatest confidence, recommend them as thoroughly fireproof.[4]

The endorsement is followed by a lengthy list of prominent Vancouver businessmen, Mayor Malcolm MacLean among them.

Blasting of the stumps has re-commenced—"the bomb continually ringing in our ears,"[5] laments the *Vancouver News*. A new wooden sidewalk along Water Street is rapidly nearing completion. A scow-load of new windows and doors arrives, having made a circuitous journey from New Westminster down the Fraser River to tidewater, around Point Grey and up Burrard Inlet. The Relief Committee puts out a request for the return of all government-issued tents "with all possible speed."[6] Tilley's Books and Stationery has re-opened on Cordova Street, with "subscriptions for all the leading newspapers, periodicals, new novels, new goods and novelties." The *Vancouver News* advertisements remind potential customers that shopping locally benefits Vancouver's recovery:

> Not Burned, but Badly Scorched! Encourage Home Industry by Spending Your Money at Home. Landreth and McCutcheon furniture and upholstery store on Abbott. Give them a call before purchasing elsewhere.[7]

> Fire! Fire! Grant and Arkell, Vancouver are again in the field with a full line of General Merchandise, such as Dry Goods, Groceries, etc., all of which will be sold at Victoria prices.[8]

Vancouver's Chinese community is re-establishing itself, with a large number of individuals taking up residence on a 160-acre tract of land along Westminster Road. The land has been leased

Cordova Street five weeks after the fire, looking west from Carrall.

to them rent-free for ten years, on the condition that they clear and cultivate it. While the re-seeding of vegetable gardens throughout the city has been slow to materialize, hoteliers and restaurant owners are clamouring for fresh produce to serve their guests. The Chinese have been hard at work grub-hoeing and clearing the Westminster Road acreage. Their market wagon, brimming with vegetables of every description, has become a welcome sight on Vancouver streets.

It takes three weeks for Vancouver's first fire engine, a John D. Ronald Company steam pumper size No. 2, to be transported by rail from Brussels, Ontario, arriving in Port Moody on the evening of Friday, July 30. A debate ensues about the best way to deliver the engine to its final home—via Burrard Inlet by barge, or over the New Road via New Westminster. Vancouver's Fire, Water and Light Committee decide to save precious dollars with the slower alternative, and commission the cartage company Berry and Rutherford to transport the engine overland by a four-horse team. John Ronald himself had arrived at Port Moody in advance of the engine to oversee personally the final leg of its journey. By 6 p.m. on the evening of August 1, Vancouver's first fire engine has arrived, and is temporarily located on a vacant lot opposite the Relief Committee headquarters on Water Street. Ronald reports that the horses handled their load with ease and that the overland journey has proved to be beneficial—providing lubrication for the engine's wheels and ensuring that everything is in good working order. The next morning, Ronald sets to work cleaning off the engine's thick coating of grease and whitewash. In an article entitled "VANCOUVER'S LEVIATHAN," the *Vancouver News* reports his progress to an increasingly curious readership:

> All day Monday and Tuesday, Mr. Ronald and his assistants were engaged in restoring the fire engine to its original good looks, the long journey in an open car across the continent having taken off the brilliancy of the brass fittings and dimmed the handsome polish with which it was covered. The engine was of too great a height to be put into a freight car; it had to be transported in an open wagon, where, during the three weeks it was exposed to the air, it received benefit of all the storms and rains which met it on the way. When the engine arrived in this city on Sunday evening its reception was a little cold. The dust through which it had passed in journeying from Port Moody via road had adhered to the grease and whitewash, with which it was covered to protect it from rust and corrosion, making its appearance anything but what one would have expected from a bright and new machine.

If Vancouverites were initially disappointed with the outward esthetics of their new fire engine, they are soon to be cajoled:

> Elbow grease and polishing material were all that were needed to take off the outside accumulation which blurred the original handsome appearance of the steamer, and it was at this work that the builder of the engine spent the whole of Monday and part of yesterday in doing so, with the help of his assistants. Very soon the beautiful brass work with which the machine is adorned began to assume its natural appearance. The brilliancy and golden hue of the fittings came out in bold relief and many were the praises which were showered upon our engine by our citizens, who crowded around the building on Water Street where it temporarily stands.

Late on the afternoon of August 3, Chief Engineer John Carlisle, along with the members of the Vancouver Hose, Hook and Ladder Company No. 1, carefully pull the machine up the newly constructed boardwalk along Water Street, to the foot of the Cambie Street wharf. Despite its five-thousand-pound weight, the engine proves to roll with ease to the water's edge. A stack of kindling and coal pieces are lighted within the engine burner and everyone anxiously awaits—some keeping a wary eye on their pocket watches. At nineteen minutes after 7 p.m., smoke appears at the top of the engine stack, soon to be followed by a burst of flames. Three minutes later, enough steam is generated to sound the engine's whistle. Four minutes following, the indicator shows that pressure has reached forty-five pounds to the inch:

> Mr. Ronald at once turned on steam and the piston rapidly rose up and down, and the work of drawing water up from the harbor commenced. Five hundred feet of hose were attached and it was not long before the sea-water had reached the nozzle and showed itself in a fine stream which was then thrown down Water street, the Stag and Pheasant receiving a good shower bath. The boys amused themselves and the spectators, by pretty freely sprinkling any citizen who advanced within reach, even his Worship the Mayor coming in for a share of this harmless pleasantry.[9]

John Carlisle's wife is given the honour of christening the new fire engine. All able hands are recruited to lift the five-thousand-pound *M.A. MacLean* (named in honour of Vancouver's mayor) off the ground, before she smashes a bottle over its metal frame.

Costs are itemized in the *Vancouver News*: Steamer—$3800, four-horse carts, ordinary price $200 apiece—$700; 2000 feet of rubber hose at $1.18 per foot—$2360, freight charges—$300, Incidentals, cost of erecting engine—$340, for a grand total of $7500. John Ronald agrees to giving ten years credit for the sum of $6,800, due to him with interest at the rate of 7 percent, per annum. Although further testing will be needed over ensuing days, everyone is in agreement that Vancouver's latest amenity, albeit expensive, is a resounding success:

Vancouver Volunteer Fire Brigade, 1887. Hugh Campbell in middle row, far left.

> The 2,000 feet of rubber hose and the four horse carriages make the equipment as perfect as it can be, and we believe that there is no city of the same size and population as Vancouver which can boast of having such a complete system of protection from fire as our town now possesses.[10]

At the next weekly meeting of the Volunteer Fire Brigade, members quickly decide that they should look, as well as act, the part of firemen—despite not having salaried positions. Two companies are formed: the Invincible Hose Company Number

1, to be identified by blue shirts; and the Reliance Hose Company Number 2, to be identified by red shirts. By August 9, the *M.A. MacLean* has been relocated to its temporary shelter—a rented Water Street building, complete with living quarters for engineer Edward T. Morris—hired by the city at sixty dollars per month. Not about to have his men idle away their hours while awaiting the first alarm, Fire Chief Sam Pedgrift orders regular practice sessions with the new engine and hose reels. Vancouver streets are given a frequent wash, much to the appreciation of the city's gentrified inhabitants—their newly sewn clothing forever at the mercy of the clouds of summertime dust being stirred up by passing wagons and carriages.

The Relief Committee efforts are winding down, and John Boultbee is asked to publish an audited report, which appears in the August 10 *Vancouver News*. Cash contributions from June 15 to August 1 have totaled $22,980.00. An itemized list of cash and in-kind contributions paid out to fire refugee claimants totals $23,179.23. To everyone on the Relief Committee's satisfaction, a relatively manageable shortfall of $199.23 is owed to city merchants.[11]

Near 11 p.m. on August 11, the Invincible and Reliance hose companies receive their first proper test when a fire breaks out at Spratt's Ark. Situated on the water at the north foot of Burrard Street, the once-bustling fish-oil refinery is now largely deserted, save for a few machine shops and nearby homes of former employees. By the time enough crewmen have been mustered to pull the fire engine up the dark, bumpy and meandering forest trail to Burrard, there is little left of Spratt's Ark to save. Not to be denied some sense of satisfaction for their efforts, the men

Vancouver Volunteer Fire Brigade hose reel team outside Fire Hall Number 1, c. 1887.

turn their hoses on adjacent buildings—a largely unnecessary action as there is no real danger of the fire spreading, but it earns some complimentary press:

> The new fire engine was a superior article and the firemen, though inexperienced, worked with a will for an hour and a half and saved an adjacent warehouse and office.[12]

A large party of Vancouverites, including Mayor MacLean, board a special convoy of stage coaches for New Westminster at

7 a.m. on the morning of Saturday, August 14. Sir John A. Macdonald and his party are arriving in the city aboard coal baron Robert Dunsmuir's steamer *Alexander* from Nanaimo, and a gala reception has been organized. Shortly after noon, the prime minister and his wife are treated to a rousing cheer of greeting from a crowd of over one thousand on the public grounds. Mayor MacLean delivers a welcoming speech, tactfully interjecting several references to Vancouver—more likely than not as a gentle reprimand that the leader of Canada has chosen to bypass the recovering city in the course of his visit to the west coast:

> SIR—On behalf of the Corporation and Citizens of Vancouver, permit me on this auspicious occasion to tender to you a hearty welcome to our young city and the Pacific Coast.

The mayor graciously acknowledges the federal government's financial contribution to the relief effort:

> Permit me here to return to you and the Government of Canada through you our heartfelt gratitude for the generous manner in which you responded to the appeal of our suffering citizens for aid in their calamity. Believe me, your assistance in our sudden distress was invaluable and will be ever remembered, and our only regret is that in consequence of our late misfortune, we are unable to give expression to our feeling in a more adequate manner, but we hope that when next you honour us with a visit we shall be in a position to do the occasion the justice which we desire.[13]

On the evening of August 24, Vancouverites are finally able to hold a social soiree of their own. A gala fundraising ball is held at the newly built Gold's Hall, near Water and Abbott to celebrate the new fire engine and raise funds for the new uniforms and other expenses. The *Vancouver News* records a vivid portrayal of the festivities:

> On entering the hall last evening a brilliant sight met the gaze. The platform was prettily decorated: a Canadian ensign, kindly lent for the occasion by Mr. Power of Moodyville, was hung at the back, and in front were two sections of hose with the shiny brass nozzles crossed. A banner inscribed with the word "WELCOME" surmounted the whole, "INVINCIBLE HOSE COMPANY NO. 1" and "RELIANCE HOSE COMPANY NO. 2" being lettered on the side. The walls of the hall were decorated with flags of all nations (lent by the captain of the German barque *Gotha*, now lying in the harbor), and numerous boughs of evergreen fir.
>
> Among the guests were Mayor and Mrs. MacLean, Mr. and Mrs. McColl of New Westminster and Assistant Chief Webb, also of New Westminster. The dance program consisted of 24 reels, waltzes, quadrilles, schottisches, as well as other dances. After the dance, at midnight, the majority adjourned to the St. Julien Restaurant for a sumptuous banquet of many various meats, salads, sauces and dessert.[14]

However elegant the evening, attendees were forced to improvise, as recalled in later years by Minnie Gardner Johnson: "It was great fun! We all wore gum boots, the roads were so muddy, and the men carried lanterns. None of us had a good dress, but we did have a good time."[15]

Greater care with safety is taken once burning is permissible again. Before the fire, CPR contractor Thomas Boyd of Boyd and Clandenning was paid twenty-six dollars an acre for the slashing and felling of District Lot 541 as far west as Burrard Street, with an additional two dollars an acre for cutting off branches. The financial return was dubious for such labour-intensive work—thus the reckless "get it down" attitude of days gone by. Now the men are paid three hundred dollars per acre to do things right—clear to the ground, and burn away every last vestige of debris without allowing progress to supersede safety.

Land speculation continues, undaunted. A giant Douglas fir, brought down at the future intersection of Granville and Georgia by the firm Stephenson and McCraney, captures the imagination of realtor James Welton Horne. It is a massive specimen, measuring fourteen feet across at the trunk and 325 feet in length. Lying at the western perimeter of the fire line, the tree has barely been singed. Douglas fir bark contains higher-than-average quantities of tannin—a natural fire retardant. Experimentally, several men try to burn through the core of the trunk's base, but only succeed in creating a small hollow—a cozy alcove, where one could shelter from the rain. James Horne jokingly fashions a publicity stunt, draping advertisements reading "Vancouver Lots for Sale" and "J.W. Horne Real Estate Office" outside the hollow. Harry Devine is there with his camera to capture the scene. The Douglas fir will eventually be carved into sections, one journeying off to England for exhibit during Queen Victoria's Golden Jubilee celebrations. In ensuing months and years, James Welton Horne's land assets will double and triple in value.

Another of Harry Devine's favourite photo subjects is the tent city hall, its whimsical "City Hall" sign still jauntily perched atop the ridgepole. The tent has served its purpose well over the summer months as Vancouver's interim location for conducting city business, but construction is underway on one of Vancouver's first buildings to be made entirely of brick—a single-storey company store for brothers David and Isaac Oppenheimer at 28 Powell Street, which will also serve as the new temporary city hall.

Not about to lose the opportunity to record yet another iconic

J.W. Horne's real estate "office," 1886.

post-fire-scene, Harry Devine asks Mayor MacLean and his officials to bring their tables and chairs outside the tent flap and arrange themselves for what will arguably become one of the most famous Vancouver historical photographs of all time. Devine leaves no minor detail unattended, even adding a vacant chair for Alderman Harry Hemlow, who is away on business in Seattle. George Gibson, in town to purchase supplies for his homestead far up Howe Sound, stops to watch the photo being taken. The future namesake of Gibson's Landing inadvertently immortalizes himself among Vancouver's decision-makers. Resplendent in new police uniforms from Seattle, Chief John Stewart, Constables Jackson Abray, Vickers Haywood and John McLaren also line up to have their picture taken.

Construction of a proper fire hall to house the *M.A. MacLean* is well underway, near the former location of the Jonathan Miller homestead at 14 Water Street. Priced at $743, it is large enough to hold the engine and other firefighting apparatus on the main floor, with sleeping quarters upstairs for twelve men. A sixty-foot-high tower, twelve-feet square, is constructed to suspend the hoses for drying and house a five-hundred-pound fire bell—purchased from the U.S. Bell Company of Troy, New York, at a cost of $150.[16]

The price of establishing a firefighting arsenal for Vancouver is rapidly climbing and matters are made no better when Fire Chief Sam Pedgrift decides to abscond with funds raised from a firefighter's charity minstrel show. Rumoured to have headed to the U.S., Chief Pedgrift, his family and the money are never to be seen again. In weeks ahead, cartage operator John Carlisle—a hard-working and respected individual, will be voted in as the new chief and will hold the position for the next forty-two years.

Fire Hall Number 1 on Water Street, with steam pumper M.A. MacLean *on right, 1895.*

On November 22, 1886, sixteen-year-old Elida Bell sits at her desk in the reopened Hastings Mill schoolhouse, pencilling out her latest writing assignment in tidy, even script:

> VANCOUVER
>
> Vancouver is a small city situated on Burrard Inlet. About a year and a half ago, there was scarcely any people there, and only a few houses. Now it has a population of nearly four thousand. About six months ago, the entire city of Vancouver was destroyed by fire started by burning brush. But in those six months it has again regained its former position as a flourishing town. It has a good harbour and a number of ships come there to load lumber for foreign lands. It is expected that Vancouver will be the terminus of the Canadian Pacific Railroad.[17]

Elida and her classmates will soon be facing more upheaval. Several buildings situated on the CPR's beachside right-of-way are slated for demolition or relocation—among them, iconic fire survivors such as the schoolhouse, the Alexander and Coldwell residences. There is no public outcry. Facilitating the arrival of the train takes precedence.

Christmas Day 1886, the *Vancouver News* paints a poignant scene of a recovered city in full celebratory mode:

> The unusually gay and animated appearance of the shops and streets last night had a tinge of the Christmas Eve spirit that most of the citizens have been accustomed to in the old settled towns in the east. Men, women and children were out in large numbers making their purchases for today's festivities, and, as they passed to and fro in the light which shone forth from the store windows, a cheerful aspect was presented to the observer. Many of the business places were gaily decorated and looked very attractive: The butcher-stalls, fruit and candy stores, stationery and fancy goods stores, and other places whose resources

are generally drawn upon for Christmas fare and presents, were liberally patronized, and scores of little folks, and old folks too, will have been made happy this morning with the pretty things that were bought last night. Several of the merchants made displays of holiday goods that would reflect credit on an older and more pretentious town, and bore evidence of the business enterprise and energy of tradespeople who have come here to assist in building up the terminal city.[18]

In the Years to Come

As Vancouver rises from the ashes, the events of June 13, 1886 fade into memory. Newly built structures of brick and stone steadily replace those of the old cedar and Douglas fir. The prestigious Byrnes Block, with its fashionable Alhambra Hotel, rises on the former site of the Deighton House. A Hudson's Bay Company store, under the management of George W. Weeks, opens its doors January 17, 1887, on Cordova Street. On the momentous date of May 23, 1887, a flower-bedecked Engine-374 pulls the first CPR transcontinental passenger train into Vancouver before a throng of cheering spectators.

Spurred on by an influx of newly arriving settlers, fire preventative measures continue. Five underground water tanks have been strategically placed throughout the city, at the intersections of Water and Carrall, Cordova and Columbia, Hastings and Abbott, Cordova and Water and Granville and Dunsmuir. Servicing the widest and remotest region of Vancouver, the Granville and Dunsmuir tank is massive—over seventy-five feet in length, thirty feet in width and ten feet deep. New, more stringent regulations on clearing fires are set to become law.

On the evening of May 30, nearly a year to the date of the Great Fire, a group of contractors clearing land west of town along Hastings Street decide to do a quick burn-off before the new regulations take effect. Through the course of the evening, a breeze blowing from the southwest picks up speed, fanning the flames beyond the crewmen's control. As they were in 1886, conditions are parched dry and a thick pall of smoke soon drifts over the city. Vancouver's firefighters are immediately called to action. They waste no time in pulling the *M.A. MacLean* to the water tank at Hastings and Abbott, laying over two thousand feet of hose and attacking the flames with a vengeance. A few small shacks have burned down, but thanks to several cleared patches of land in the area, the fire is not spreading at a devastating rate.

Wary at the prospect of history repeating itself, Mayor MacLean and his associates quickly decide that it would be prudent to have additional fire-fighting equipment on stand-by. A telegram is dispatched to the New Westminster fire department, asking for the use of their engine. At 10 p.m., a reply is received:

> Mayor McLean: Engine will leave here in a few minutes. Will send you all the hose we have. W.F. Fagan.

Fifteen minutes later, two more telegrams arrive:

> Mayor McLean: Steam engine and fifteen hundred feet of hose ready to send to you. R. Dickenson, Mayor, New Westminster.

Mayor McLean: Engine and men have left. Have coal ready for engine. W.F. Fagan.[1]

While some Vancouverites nervously pack their valuables, the volunteer firemen continue their assault. When the flames are beyond reach of the hose, the *M.A. MacLean* is pulled to the outermost water tank at Granville and Dunsmuir. The wind has veered away from the city, causing the fire to burn back over its previous path—now devoid of fresh fuel—and providing the crewmen with a welcome opportunity to gain the upper hand by liberally soaking the area. When the New Westminster brigade arrives at 2:20 a.m., there is little left for them to do—save rest and enjoy a late night supper at Dougall House. The brush fire of 1887 eludes much notoriety, thanks to a combination of quick action by the firemen, less volume of slash and a cooperative wind. The brief story of its occurrence is relegated to page four of the May 31 *Daily News-Advertiser*.

By August of 1888, the first submerged water main, twelve inches in diameter, has been successfully installed across the Narrows to connect Vancouver with the bountiful supply of fresh water from the Capilano River. On March 26, 1889, fresh, clean drinking water flows from Vancouver taps. Cobblestone pavement eradicates the age-old problem of mud-laden potholes in the busiest thoroughfares. An opera house, several multi-storey hotels, banks, a central post office, library and hospital add new ambiance to the cityscape. Charlie Woodward opens his first dry goods store near Main and Hastings streets. Woodward's Department Store will become a legendary fixture for downtown shoppers.

Those who were "here before the Fire,"[2] becomes a benchmark of pride for longtime Vancouver citizens. In 1893, the Vancouver Pioneers Association is founded as a social organization for those families and individuals who arrived in Vancouver prior to June 13, 1886. Over ensuing years, the Pioneers will hold Stanley Park reunions, weekend sojourns across the Strait to Newcastle Island, gala parties and commemorative gatherings—keeping alive the spirit of Vancouver's earliest beginnings.

On November 3, 1898, a young Welshman by the name of James Skitt Matthews arrives in town with his new bride Valentia Maud. Passionately fond of history and journalism, he soon begins to chronicle the bygone years of his adopted city, meticulously recording and publishing the details of his findings. Although his career pursuits deviate widely over the years—from timber clerk, Imperial Oil salesman, and decorated military major to entrepreneur, scow and tugboat operator, Major Matthews' fascination with Vancouver history remains constant. He will personally amass volumes of documentation, compile forty publications and interview dozens of early Vancouver citizens from all walks of life, with a passion bordering on the obsessive. In particular, recollections of the Great Vancouver Fire capture his imagination. Raised in the small settlement of Te Horo, New Zealand, young James

Commemorative ribbon.

witnessed at an early age a grass fire consume a large tract of ranchland surrounding his boyhood home —an event which resulted in the destruction of his parent's livelihood and left him with a profound empathy for those who had suffered similar hardship.

Many of Major Matthews' interviews are with rapidly aging Vancouver pioneers, whose often fleeting thoughts and diminishing memories result in transcribed pages that dart randomly from one topic to the next. The Major tries to resolve the problem by having the predominant subject matter of each paragraph typed in upper case and underlined. "GREAT FIRE 1886" entries are liberally sprinkled throughout his recorded text. Some pioneers, such as Mrs. Emily (Strathie) Eldon and Annie Sanders Ramage, contribute their own Great Fire reports, full of graphic, eye-witness details. Emily Eldon entitles her contribution, EARLY DAYS IN VANCOUVER:

> Manuscript written by Mrs. Emily Eldon, 1150 Alberni Street, at the request (of) Major Matthews following conversation at Pioneers Picnic, Newcastle Island, 15 June, 1932.
>
> The one redeeming feature of the Great Fire, the anniversary of which pioneers celebrate each year, is that it is impressed indelibly upon the minds of those who lived through it, conditions as they existed in that part of the city's history. Other days may have vanished from our minds, but memories of that event, of Vancouver, its environs, and the people of that day, are as clear now as they were a week after the fire.[3]

Annie Sanders Ramage contributes "A DESCRIPTION OF THE GREAT FIRE, 13th June, 1886, by a girl who passed through it."

> By Mrs. S.H. Ramage, 1110 West Eighth Ave., Vancouver, "Princess Anne", daughter of Mr. and Mrs. Edwin Sanders, pioneers, "Before the Fire, who came here in March, 1886—before

> the railway reached Port Moody. Their home was one of the few which escaped the "Great Fire". Mrs. Ramage is the sole survivor of the famous "Coffee Brigade" of Vancouver, a group of pioneer women who followed the Vancouver Volunteer Fire Brigade to fires, and supplied them, whilst the men fought fire, with hot coffee.[4]

Most of those individuals who bore witness to the Great Fire are deeply moved by the experience. A few, like early Vancouver pioneer Harry Hooper—interviewed by Major Matthews during a return visit to the city from his placer mine on the banks of the Fraser River—are less than impressed:

> There is a lot of bosh talked about the Great Fire; the pioneers did not suffer so very much. It was summertime; you could sleep out under the trees, and the fire did not go past Hastings Mill much; the fire missed a lot of that locality.[5]

Harry Hooper, a young lad of six at the time of the fire, had clearly thought of the experience as one grand adventure.

Fascinated with the sudden and destructive nature of the fire, Major Matthews sketches a cartoon-style map of Vancouver and surrounding environs as they appeared on the afternoon of June 13, 1886, based on eye-witness descriptions. A.E. White of the Art Engraving Company, 532 Pender Street, is commissioned to provide the professional finishing touches with India ink and water colours. The result is a graphic depiction of a city under siege, with fiery red flames taking dead aim at the very heart of Vancouver's most built-up city streets. George Schetky, of the 1886 Vancouver Fire Brigade, and Thomas Mathews provide their signatures to verify the map's authenticity. Major Matthews' Great Vancouver Fire Map[6] is published in the June 12, 1932, Sunday Magazine section of the *Vancouver Daily Province* and will be reproduced many times over ensuing years.

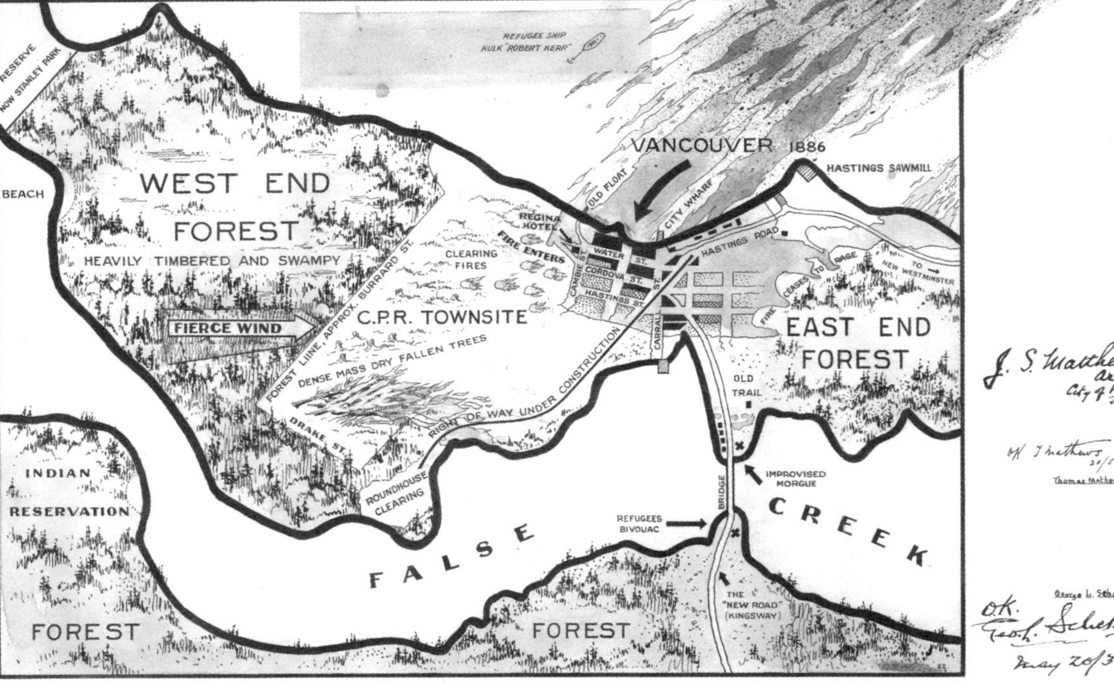

The Great Vancouver Fire map.

An eclectic assortment of Great Fire artifacts are offered up to the Major's care, their owners anxious to see them preserved for the interest of future generations: Emily Strathie Eldon's pocket watch, the Reid family's iron cooking pot, the melted bell of St. James Church, melted shot glasses from the Sunnyside Hotel, Anna McNeil's melted jewellery, a checkered blouse worn by seven-year-old Greta Miller as she fled with her mother and siblings to Hastings Mill, a set of keys from Otton and McGregor's grocery store. Major Matthews accepts each item with reverence, attaching neatly handwritten histories, or engraved commemorative plaques whenever possible. Every donor receives a written acknowledgement that often goes well beyond the typical thank-you letter. The one received by Margaret Stevenson, (formerly Margaret "Greta" Miller) in April of 1957 offers a good example:

> Dear Mrs. Stevenson:
>
> Whether I missed you, or not, at the "HERE BEFORE THE TRAIN" dinner on Monday evening I do not know, but I do not recall seeing you there. Everything seemed to go alright, and some of them lingered until after 9 p.m.
>
> This morning I have enclosed the child's blouse in a frame with glass, but as a temporary trial only, to see how it looks, taking care to have the embroidery of the collar and sleeves exposed to the full—they are pretty. The people who lived in Granville were not a lot of savages, as some of those writers would make them out to be. And, I suppose there are those who would take me for an old jackass who "sees things". Well, what I see in this blouse is a little bit of the life of Vancouver before it was Vancouver, and, if that is not history, I should like to know what is.
>
> I presume to enclose, under the glass, an inscription something like this:—*This child's blouse* was worn by Miss Margaret Miller "Greta" Born Granville, Burrard Inlet Sept. 1879, daughter of

Jonathan Miller, first postmaster of Vancouver, as she fled from the Great Fire which destroyed the first Vancouver, 13th June 1886.

If convenient, would you please phone me, or Mrs. Gibbs, if this inscription is correct. Then I shall proceed to inscribe it, and seal the frame up dust and dirt proof. With my deep respects.

Most sincerely,
Major J.S. Matthews, City Archivist.[7]

Vancouver's first fire engine, the *M.A. MacLean* steam pumper, rapidly becomes obsolete with Vancouver's growing population and the advent of faster, more efficient machines. With a cracked boiler and in a poor state of repair, it is sold for scrap metal in 1905. Around the same year, the Regina Hotel is demolished to make way for the four-storey, stone-clad Edward Hotel at the southwest corner of Cambie and Water streets. A July 1, 1906, liquor licensing bylaw prohibits the sale of alcoholic beverages in any establishment other than a hotel bar or restaurant. The glory years of Vancouver's rustic, independently-operated saloons are a thing of the past.

The *Robert Kerr*, saviour ship of Vancouver, meets an unceremonious end. Having been sold by Captain Soule to the CPR, the vessel serves as a coal hulk for many years, transporting coal to replenish the various steam ships plying local waters. On March 4, 1911, the *Robert Kerr* is under tow by the steam tug *Coulti*, transporting 1,800 tons of coal destined for the CPR steamship *Empress of India*, moored in Vancouver. During the night, the *Coulti* veers off course, causing the *Robert Kerr*'s bow to strike an uncharted reef off the north end of Thetis Island. Unlike the San Juan Island grounding of 1885, this incident sounds the

Blouse worn by Greta Miller as she ran from the fire.

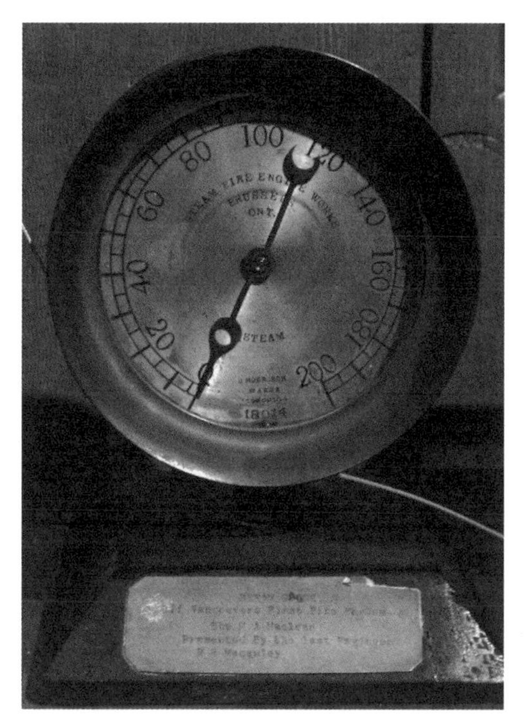

Steam gauge from the M.A. MacLean *steam pumper.*

death knell—the vessel quickly fills with water and sinks to the bottom. Two days later, a diver confirms that the *Robert Kerr*, lying on its starboard side in twenty to sixty feet of water, is a total loss. The wreck is sold at auction to the Vancouver Dredge and Salvage Company, whose crewmen manage to reclaim a profitable portion of its coal cargo before abandoning their efforts. At some point in time, the bell of the *Robert Kerr* is brought to the surface, and continues to be rung every April, at the Vancouver Historical Society's annual Incorporation Day Luncheon.

On June 13, 1925, thirty-nine years after the fire, a memorial drinking fountain is unveiled before a large crowd at Carrall and Water streets by early Vancouver pioneer Annabella Fraser, wife of logging operator Angus Fraser. (It was originally planned for Margaret MacLean, the widow of Vancouver's first mayor, to perform the unveiling, but she had been forced to cancel due to illness.) Known as the "Maple Tree Memorial," the fountain features a bronze tableau sculpted by Charles Marega, of a small gathering of pioneers beneath the leafy boughs of the grand old tree that formerly occupied the square. Charles Gardner Johnson and Pete Righter—engineer of the first CPR train to arrive in Vancouver, are given the honour of taking the first drinks from the bubbling cascade of water. Mayor Lewis Denison Taylor, a strong advocate for the preservation of local history, formally declares that June 13 will be henceforth known as "Vancouver Day"—a time to reflect on the early Vancouver pioneers and their years of hard labour and self-sacrifice in bringing their city to prominence.[8] The significance of June 13 is twofold. In addition to being the anniversary of the Great Fire, it is also the date that Captain George Vancouver and his crewmen explored the waters of Burrard Inlet during their 1792 west-coast survey.

Pioneers gathered for unveiling of memorial fountain, June 13, 1925.

(The fountain is now gone, but Charles Marega's original tableau can still be found, affixed beneath the statue of John "Gassy Jack" Deighton in Maple Tree Square.)

By far the largest, oldest and arguably most intriguing Great Fire "artifact" is the old Hastings Mill Store. By the late 1920s, the Hastings Sawmill had been closed and the weathered buildings were being demolished to make way for up-to-date port facilities. Native Daughters of B.C. Post #1, a 1919-formed society of historically minded, born-in-B.C. ladies, rallied to the cause of Vancouver's oldest surviving building. Mill owner Eric Hamber, husband of Aldyen Irene (whose father John Hendry had managed the company for many years), agrees to offer up the store free-of-charge—on the condition that it be moved off the premises at the Daughter's expense. On the evening of July 24, 1930, an emergency meeting of Native Daughters Post #1 is held in the Women's Building at 752 Thurlow Street to discuss their strategy. It is a daunting prospect, but the resolute Daughters undertake a whirlwind fundraising effort.

On July 28, 1930, under the watchful eye of several Native Daughters aboard the escort vessel *Fispa*, Hastings Mill Store is lowered onto a scow and towed out of Burrard Inlet by Captain Charles Cates aboard the tugboat *Alert*. Making good use of the evening full tide, Cates completes the tow across English Bay and beaches the scow beneath a forty-foot escarpment at the north end of Alma Road. In the daylight hours, crowds gather to watch and cheer as the store is painstakingly winched up the escarpment to its new foundation. House-mover Frederick Gosse cannot contain his fascination with the task at hand. "It's standing the strain well, what stuff they put into that old building—real lumber! And how well those pioneers did their work. I don't find many buildings so staunch and strong after so many years!"[9]

The ladies continue with their fundraising drive, bolstered by regular coverage and published lists of contributors in the *Vancouver Sun* and *Vancouver Province* newspapers. Jessie Greer Hall, Great Vancouver Fire survivor and stalwart Past Chief Factor of the Native Daughters Post #1, pens a letter of appeal:

> That the Native Daughters of B.C. are the instigators of the great campaign to preserve our local history should be known to our citizens. Their earnest support and cooperation is urged, for in so doing, they are helping to build up the national history of this province and the Dominion, not only to their own great pride, but for the glory of their children in the generation to come. Wake up, Vancouver![10]

Whether or not Jessie's letter stirs the city from its Depression-era slumber is initially questionable. After two weeks of intensive fundraising, the Daughters manage to raise nearly five

Hastings Mill Store being barged to its new home.

thousand dollars—but that is only half of the estimated total cost to relocate and renovate the store. Much to everyone's surprise and pleasure, a wealth of in-kind donations help to bring the project to completion over ensuing months. Captain Charles Cates waives the $89.00 towing charge. False Creek Lumber, B.C. Hardwood Floors, Sherwin-Williams Paint, Dominion Bridge Construction, Victoria Brick and Tile, and Jarvis Electric are among the many businesses who contribute labour and materials.[11] B.C. Chamber of Mines arranges for the addition of a splendid fireplace of B.C. granite. On January 10, 1931, Hastings Mill Store, sole remaining structure to pre-date the Great Vancouver Fire, is officially opened by Lieutenant Governor Randolph Bruce. A year later, it will be dedicated as a Museum of Historic Relics by Premier Simon Fraser Tolmie and the Honourable Nelson Seymour Lougheed, Minister of Lands.

Old Hastings Mill Store Museum, August, 1932.

Confusion surrounds the eventual fate of the old Gilliland telephone exchange rescued from the fire by Seth and Charley Tilley. A telephone exchange displayed in the lobby of the B.C. Telephone building on Seymour Street is said to be Vancouver's first telephone—the very same Gilliland that was so heroically saved. No one, however, seems to know exactly when or how it arrived there. Eager to track down the full story, Major Matthews interviews George Hodge, former manager of the Burrard Inlet and New Westminster Telephone Company, one January day in 1934. During his recollections, George casts a heavy shadow of doubt on the exchange's authenticity:

> No, the *first* switchboard in Vancouver is not at the B.C. Telephone Company's office. I think you'll find the *first* switchboard in Vancouver in the middle of Burrard Inlet. The one the B.C. Telephone has was the Moodyville switchboard. The first Vancouver switchboard was dumped into the center of the inlet, about 1900, together with a very valuable lot, for historical purposes, of old telephones and old equipment, cleaned out of the bottom of the basement of the old Lefevre Block, northwest corner of Seymour and Hastings streets. I am almost sure that in that bunch was the old Gilliland Board . . .[12]

Somewhat disappointed with this unexpected verdict, Major Matthews decides to seek out the one individual who will be able to, in his mind, properly identify the exchange—George Pittendrigh, who installed it at Tilley's Books and Stationery so many years ago. In 1938, George Pittendrigh is photographed alongside the exchange, reasonably certain that it is the same piece of equipment that he was hired to hook up at Tilley's. After that, the old exchange simply vanishes into oblivion. Its whereabouts remains a mystery today.

George Pittendrigh, with Tilley's alleged telephone exchange.

Great Fire relics of the natural kind—partially burned logs, stumps, the damaged Princess Louise Tree, the famous old maple of Maple Tree Square—were cleared away with rapid dispatch in the days following the fire, with one notable exception. A solitary charred stump, somehow left in place alongside the CPR right-of-way near Carrall and Cordova, is finally removed on August 8, 1939, to clear space for an Army and Navy department store parking lot. There is little fanfare for the old survivor, which was estimated to be two hundred years of age at the time of the fire—its memory immortalized in the famous Harry Devine tent photo, according to recollections of Walter Graveley.

Periodically, outdated public documents regarded to be of no general value are approved for destruction by the printing committee of the Provincial Legislature. In 1951, a heap of paper-

Removal of the last stump, August 8, 1939.

work designated for the furnace is randomly perused by provincial archivist Willard E. Ireland. He is shocked and horrified to discover, amidst fragile sheets crumbling with age, the original Great Vancouver Fire coroner's report of June 14, 1886. A stop work order is instantly placed on the entire collection. No one can explain how such a valuable and historic document became earmarked for rubbish. The prominent B.C. historian and journalist, Bruce Alistair McKelvie, publishes details of its remarkable background and resurrection in the September 22, 1951 magazine section of the *Vancouver Daily Province*.

Vancouver Day comes and goes over many years with little acknowledgement, until the city is one hundred years old. On June 13, 1986, Vancouver is in the midst of hosting the Expo '86 world's fair, a giant, five-month-long centennial celebration flanking the north and eastern shores of False Creek. Coincidentally, it is the hottest June 13 on record in the city, with a daytime high of 25.6 degrees C.[13] At 2 p.m.—the same time the Great Vancouver Fire bore down on an unsuspecting populace one hundred years ago, alarms are sounded on old and new fire engines on display in Gastown, as well as in fire halls throughout the city. The same day, medallions commemorating the Great Vancouver Fire are distributed to Vancouver elementary schoolchildren.

On June 12, 2011, during year-long festivities to celebrate Vancouver's 125th birthday, the Vancouver Fire Fighter's Historical Society and Vancouver Fire and Rescue Services team up to commemorate the Great Vancouver Fire in fine style. A parade of vintage fire trucks makes its way through the streets of Gastown, the Vancouver Fire and Rescue Services band pro-

Great Vancouver Fire 100th anniversary commemorative medal.

vides musical entertainment and Vancouver historian John Atkin—well known for his interpretive walking tours, describes events as they occurred on June 13, 1886 to a rapt audience. Representatives from Native Daughters of B.C. Post #1 proudly display the 1886 coffee urn, making its first re-appearance near to the location where it was used to dispense hot coffee to fire survivors so many years ago.

Beneath the rebuilt city structures, demolitions and evolving skyline of Vancouver's downtown core, there still exists tangible evidence of the events that occurred on June 13, 1886—a layer of Great Vancouver Fire ash, which has received little mention over ensuing years. Excavation crews frequently encounter the layer while bulldozing through the subterranean soil of many a downtown construction site. During a 2007 renovation project at the Gospel Mission, located on the west side of Carrall Street midway between Cordova and Hastings, ashes are discovered beneath the basement floor. Gospel Mission occupies a building formerly known as the Louvre Hotel. The Louvre, originally constructed in 1889 as the Vancouver Tea and Coffee Company, had an earth floor basement, where the ash lay undisturbed for decades before being discovered.

Today, Vancouver is a vibrant, cosmopolitan city, consistently ranked among the most livable places on earth. Locals and visitors browse in the gift stores and dine in the trendy bars and restaurants that line Water Street where once the fire raged. Glistening Alaska-bound cruise ships anchor dockside near where Emily and Alexander Strathie struggled to stay afloat aboard their makeshift raft. Stacked cargo containers and towering gantries line the dockyards where Hastings Mill employees beat at rampaging flames with blankets. The Carrall Street Greenway project is taking shape where William Gallagher hurried to gather belongings from his office before it burst into flames. Recording artists fine-tune their music in "The Warehouse Studio," renowned singer Bryan Adams' ambitious renovation of the Oppenheimer brother's 1886 cement and brick warehouse. Glass towers of the 2010 Winter Olympic Village are clustered where fire refugees gathered on the south shore of False Creek, anxiously awaiting the supply wagons from New Westminster.

It is hard to believe that on June 13, 1886, the thriving, 21st-century metropolis of Vancouver was indeed—a city of ashes.

Whatever Became of...?

Whether or not those individuals who witnessed the Great Vancouver Fire harboured any personal demons or post-traumatic stress is open to conjecture. Over the years, some faded into obscurity—the documented trail of their lives running cold. Others carried on with their day-to-day activities, engaging in productive careers and community service, raising families and leaving their own colourful stamp on Vancouver.

After serving for two years as Vancouver's first mayor, Malcolm Alexander MacLean moved on to become a Police Magistrate and Justice of the Peace, gaining wide respect for his fair and judicial decisions. He became president of the Vancouver Pioneer Association and also helped to inaugurate the Vancouver St. Andrew's and Caledonian Society—a social group for

Vancouverites of Scottish ancestry. Eternally proud of his adopted city, MacLean actively encouraged immigration, particularly among his Scottish countrymen. After his untimely death on April 4, 1895, at fifty years of age, thousands of Vancouver citizens stood solemnly in tribute as Malcolm MacLean's funeral cortege wound through the downtown streets. New MacLean Park in the Vancouver neighbourhood of Strathcona commemorates his memory.

CPR Western Division superintendent Henry (Harry) Abbott continued to be involved in the growth and modernization of Vancouver. He oversaw establishment of the Vancouver Electric Illuminating Company, which powered electric lights for fifty-three city homes and three hundred street lamps on its inaugural day, August 8, 1887. City planners drew heavily upon his railway expertise while developing Vancouver's first streetcar system—a 5.4 kilometre network spanning several major downtown thoroughfares. Local marine transport also flourished under Henry Abbott's watch, with increased steamship service between Vancouver, Victoria and Seattle. Henry Abbott retired from the CPR in 1897, and died on September 13, 1915 at eighty-six years of age. Vancouver's Abbott Street, the Fraser Valley community of Abbotsford and a mountain in the Selkirk Range are named in his honour.

John Joseph Blake became Vancouver's first city solicitor and compiled the first Vancouver City Charter—establishing the legal criteria for Vancouver's operating budget, by-law preparation and other aspects of city governance. He married Elizabeth Ann Fawcett, the widow of Great Fire victim Albert Fawcett on July 25, 1887. John Blake died on February 6, 1899.

Alan McCartney, trained as a surveyor, civil engineer and architect, worked on projects in both Vancouver and North Vancouver. On December 29, 1890, he attended a meeting to discuss the formation of a municipality on the North Shore. In later months, McCartney would be instructed to draw a map of the new municipality to accompany a petition to the Provincial Legislature. He became a councilor for the Corporation of the District of North Vancouver in 1892 and surveyed routes for the future Mount Seymour Parkway and Dollarton Highway. Alan McCartney died in May of 1901 at the age of forty-nine. North Vancouver's McCartney Creek is named in his honour.

John Boultbee, Vancouver's first magistrate, re-opened his law office in the new Ferguson Block at Powell and Carrall streets. He remained in Vancouver for another twelve years, before moving to Rossland, B.C., where he partnered with John Rankin in mining, stock brokerage, conveyors and notaries. Often in talking of his Great Fire experience, John would point to three permanently bald patches in his scalp where hot cinders landed while he lay in the hollow. John Boultbee died in Vernon, B.C., on August 23, 1906 at fifty-four years of age.

Dr. William McGuigan continued his practice at the southeast corner of Cordova and Abbott for many years, in partnership with Dr. Henri Langis. Becoming known as the "most titled man in Vancouver,"[1] Dr. McGuigan served as City Coroner, alderman, health inspector for the city's First Nation population, secretary of the B.C. Medical Association, chairman of the Free Library Board, and president of the High School Board, among numerous other positions over the years. The good doctor presided as mayor of Vancouver in 1904, a term during which he saw the conversion of east False Creek from tidal mudflats to CNR rail yards. Dr. McGuigan gradually declined in health and lived his final two years at St. Paul's Hospital (which he had helped to found), before passing away on Christmas Day 1908, at the age of fifty-five. He is buried at Mountain View Cemetery alongside his lifelong friend, John Blake.

Thomas McGuigan, brother of William, died on June 26, 1910, one day before his fiftieth birthday, after a long and distinguished career as Vancouver's city clerk. Well respected by all, Thomas was given a fitting obituary, published in the June 27 *Vancouver World*. "He carried on the work of the city accurately and at a time when a huge amount of the worry fell on the shoulders of the City Clerk."[2] The original city minutes that Thomas McGuigan rescued from the Great Fire continue to be maintained in Vancouver City Archives, in pristine condition.

Despite losing an estimated eight thousand dollars' worth of stock, Seth Tilley quickly re-established his books and stationery business. He and his wife moved to Bella Bella in the late 1890s, where they managed a store and post office. Seth Tilley died on August 21, 1910, at seventy-five years of age. Charley Tilley worked as manager of the telephone exchange and then as a purser on the *Empress of India* until his death in 1898.

Father Clinton saw the reconstruction of his St. James Church at the corner of Gore and Cordova streets, well removed from the CPR right-of-way. He served his surrounding community with unselfish devotion, ministering to the sick and downtrodden, as well as volunteering with the fire brigade. Late in 1911, after years of hard work under often trying circumstances, he found his health declining. Accompanied by trained nurses, he set out for California with hope that a few weeks in the warm sunshine would provide restorative powers, but the cure was not to be. Father Clinton died in Paso Robles on January 29, 1912, two days before his fifty-eighth birthday. The third St. James Church to occupy Gore and Cordova streets commemorates Father Clinton's work with an altar monument.

Jonathan Miller and his family moved into a large new house on Burrard Street—then later to an estate-like property in the neighbourhood of Fairview, where Miller could indulge his passion for breeding and training race horses. He became extensively involved with Vancouver business interests, including the Vancouver Electric Illuminating Company and the Dunn-Miller Block in partnership with Thomas Dunn. Energetic and hardworking, he continued to serve as city postmaster until his retirement in 1909. After some years spent in Long Beach, California, as a result of failing health, he suffered a massive stroke. Two of his children rushed to his bedside, where he weakly expressed his fondest desire—to see his beloved Vancouver once more. With the assistance of a nurse, his wishes were fulfilled. Jonathan Miller died in Vancouver General Hospital on December 6, 1914, at the age of eighty-two.

When a new police station was built on Powell Street, John Clough happily settled in as self-proclaimed jailer, janitor and caretaker. He was given the additional job of lighting the city's gas lamps each evening and extinguishing them in the morning, although this position was to be short-lived as the ever-evolving city rapidly progressed to electric street lighting. In 1904, yet another new police station was opened and seventy-four-year-old Clough was politely asked to retire. In appreciation for his long years of service, he was given special permission from the Board of Police Commissioners to continue residing in the old Powell Street station—a welcome way to stretch his meagre monthly pension of thirty dollars. In 1911, the badly dilapidated building was slated for demolition, and John decided that it was time to return home to his English birthplace of Alford, Lincolnshire. John Clough died in the winter of 1914, at approximately eighty-four years of age.

Putting his failed Vancouver mayoralty bid behind him, Richard Alexander was elected to city council as alderman in 1887 and served until 1889. Under his ongoing direction, Hastings Sawmill grew into a major commercial hub, amalgamating with the Royal City Planing Mills, to become the British Columbia Mills, Timber and Trading Company—the largest lumber company in the province. By 1891, mill saws were churning out up to 160,000 board feet of lumber per ten-hour shift.[3] Alexander travelled extensively in Europe and South America to ensure the stability of foreign markets for B.C. wood. Active in numerous business and charitable organizations throughout his remaining years, Richard Alexander died at the age of seventy on January 29, 1915, while visiting a son in Seattle.

Despite heavy losses, Thomas Dunn quickly re-established a large new hardware store on Cordova Street, guaranteeing customers "a better selected stock than any other store in the province."[4] He formed a business partnership with Jonathan Miller, opening the Dunn-Miller Block (today's Army and Navy building) in 1889. Another venture, the Dunn Block, built on Granville and Pender streets in 1893, housed the offices of the Union Steamship Company. Thomas Dunn was heavily involved in Vancouver commercial interests as founding vice-president of the Board of Trade and president of the Vancouver Electric Illuminating Company. For the sake of his wife's health, he moved to San Diego in 1912, leaving his eldest son in charge of the Dunn business empire. Thomas Dunn died late in 1915 at the age of sixty-three. The Dunn family home—reconstructed at the northeast corner of Dunlevy and East Cordova streets in the months immediately after the Great Fire, still stands, and is reputed to be Vancouver's oldest-surviving family dwelling.

Following the Great Fire, Emma Alexander led the first

informal Vancouver Hospital Auxiliary—a milestone in what would become her lifelong passion for helping those in distress. She was a founding member of the Victorian Order of Nurses and involved with various other charities. Her knowledge of both conventional medicines and home remedies earned her the trust of many, including members of the native community. Often, an individual prescribed with medication from Dr. Bell-Irving or Dr. Walkem, would show it to Emma "for inspection and approval" before taking it.[5] Lengthy tributes to Emma Alexander were published in Vancouver's four daily newspapers, after her death on June 3, 1916, at the age of seventy-one.

Captain William Soule, master stevedore of Hastings Mill, lived with his family aboard the *Robert Kerr* for a year following the Great Fire, before eventually moving into a new home at Powell and Dunlevy streets. He worked at the mill for a total of thirty-four years—a period of time which saw enormous growth in shipment of Burrard Inlet lumber to a wide range of Pacific Rim ports-of-call. Lucrative markets expanded in Chile, Peru, Australia and China during Captain Soule's tenure. He was also heavily involved in the reconstruction of St. James and other Anglican churches in Vancouver. Captain Soule died on June 14, 1916, at the age of eighty-three. Theresa Soule, "the muscular Christian," died on March 14, 1923, at the age of seventy-eight.

Joe Fortes, shoeblack and porter of the Sunnyside Hotel, re-established himself with a bartending position at the Bodega Saloon and subsequently, the Alhambra Hotel. During his off-duty hours, Joe made regular visits to English Bay beach where he gained great popularity as a volunteer lifeguard and swimming instructor. Bowing to public demand, Vancouver City Council voted to hire Joe as an official city employee in 1900. Over the entire course of his years at English Bay, Joe taught three generations of Vancouver children how to swim and saved dozens of lives. After Joe's death on February 4, 1922, at the approximate age of fifty-seven, there was an outpouring of grief and thousands of Vancouverites lined the streets for his civic funeral. A memorial fountain in Alexandra Park, a West End library and a restaurant pay tribute to the memory of Joe Fortes—one of Vancouver's most celebrated citizens.

Lewis Carter rebuilt Carter House on its original Water Street location. The new hotel, once again three storeys in height, was a handsome structure featuring an ornate second floor balcony and large, ground floor picture windows. On November 2, 1886, Carter married Margaret Jane McMorran in New Westminster. Together they ran Carter House, advertising meals and beds for the price of twenty-five cents. Upon their retirement, Carter's younger brother John, who had been working as hotel clerk, took over the establishment. Lewis Carter died on February 20, 1923 at the age of seventy-nine.

Charles Gardner Johnson fully recovered from his injuries and participated actively in many Vancouver team sports such as cricket and rugby. His shipping agency, Gardner Johnson Limited, rose to prominence and earned Charles the title "father of Vancouver's shipping industry."[6] Active in the military for many years, he retired from the Sixth Regiment, Duke of Connaught's Own Rifles in 1911 with the rank of major. One evening at a Vancouver Pioneer Association meeting, Charles related the story of his encounter with the exploding valise and jokingly remarked, "I'd like to know who put that valise there, I have something to say to him!"[7] He laughed when informed that it belonged to Jack Alcock. Charles died on November 19, 1926, at the age of sixty-nine. Minnie Gardner Johnson died at age ninety-two on February 11, 1953.

Christina and Duncan Reid celebrated the birth of their son Alexander Campbell Reid, first white baby to be born after the Great Vancouver Fire, on July 30, 1886. The family operated a small grocery store on Cordova Street, and Duncan later became one of the city's first three letter carriers. Christina achieved a remarkable feat, being elected as Vancouver's first woman trustee to the Vancouver School Board in 1898—a time when few women dared to entertain a desire to "step out of the kitchen." She was a strong advocate of teaching science in the classroom and worked hard to affiliate Vancouver's first high school, King Edward, with Montreal's prestigious McGill University. Duncan Reid died on February 10, 1920 at the age of sixty-eight. Christina died on April 11, 1933, at the age of eighty-five.

George Allan opened his own shoe and boot business on Cordova Street after Sam Pedgrift left town with all the cash proceeds for volunteer fire-department uniforms. On speculation, he purchased property on Hastings Street, which would later become the site of the CPR Telegraph Office. Selling his land at a substantial profit, Allan travelled to Atlin, B.C. in 1898, where he managed a general store for two years before returning to Vancouver. A member of the Vancouver Pioneer Association, he continued in retail sales until his retirement at age sixty-five. A year later, Allan decided that he was not yet ready for retirement and operated a shoe repair shop on West 10th Avenue. George Allan died at his Pender Street home on Christmas Eve, 1933, at the age of seventy-five.

Frank Hart quickly re-established his furniture store and (in a move unrelated to the fire) began a side business in the manufacture of coffins. As Vancouver's first undertaker, he is said to have worn the first silk top hat in the city. Hart remained prominent in many Vancouver business and charitable circles. He provided financing for Hart's Opera House, a popular, albeit rudimentary affair on Carrall Street. The opera house—originally a Port Moody roller skating rink that had been dismantled and reconstructed in Vancouver—was a venue for dances and shows, as well as a meeting hall for the Salvation Army. In 1897, Hart joined the stampede to the Klondike goldfields, although like most of the starry-eyed individuals who ventured north, he found little success. Real estate investments in the Alaskan boomtown of Dyea proved equally disastrous, when it became known that the Yukon White Pass rail line would bypass the community in favour of nearby Skagway. Frank Hart eventually went back into the furniture business, settling in Prince Rupert, where he died on May 4, 1935, at the age of seventy-eight.

James Ross continued as editor of the *Vancouver News*, but without his partner Harkness, who sold out his interest and left for California shortly after the fire. Ross was a great booster of Vancouver and frequently printed glowing tributes to the city along with the daily news stories and advertisements. A daily weather report, not always known to be accurate, was an additional feature. Financial difficulties spelled the demise of the *Vancouver News*, its final issue appearing March 30, 1887. Ross sold out to Francis Carter-Cotton of the *Daily Advertiser* and moved back to Ontario, where he continued newspaper editing for a number of years with the *Smith's Falls Independent* and the *Winchester Press*. He made periodic visits back to Vancouver, often expressing his desire to live there again. James Ross died in Winchester, Ontario, on December 18, 1935.

Captain Thomas Jackman decided to erect his own hotel on Powell Street, rather than work in the reconstructed Sunnyside. Soon the sea beckoned again and he spent several years working with the Hudson's Bay Company as a steamboat captain on the

Skeena and Stikine rivers. He travelled to China and then quickly returned to heed the call of the 1898 Klondike gold rush. During the First World War, Captain Jackman was stationed in Victoria, helping to build ships for the French government. Through all these years of extensive travel and adventure, he often thought of Vancouver and returned to the city in time for the city's Golden Jubilee celebrations in 1936. Captain Jackman died in Vancouver on March 30, 1937.

Harry Devine assembled a remarkable collection of pre and post-fire Vancouver photographs, despite losing $800 of studio equipment. Before summer's end, his business was firmly re-established in the new Horne Block on Cordova Street, the first building to use bricks in its construction. With Vancouver's street network expanding, Devine took to bicycling from one photography project to the next, his equipment carried in a basket strapped to the handlebars, tripod carefully strapped lengthwise to the bicycle frame. He photographed many of Vancouver's historic events, such as the arrival of the first Canadian Pacific Railway passenger train and the city's 1887 Dominion Day celebrations. He is also credited with taking the last known photograph of the historic Hudson's Bay steamer *Beaver*, before it was wrecked off Prospect Point. While never entirely abandoning photography, Devine chose to join his father's more lucrative business dealings in real estate and insurance, later expanding them to include customs brokerage and warehousing. Harry Devine died on December 17, 1938, at seventy-three years of age.

Walter Graveley re-established his real estate business in a brick building located on the original triangular piece of land he had purchased from the Canadian Pacific Railway, bounded by Carrall, Cordova and the CPR tracks. Immensely proud of the fact that he was among the very first purchasers of Vancouver real estate, Graveley carefully preserved his copy of Deed Number 1 and retained the downtown property throughout his life—despite its huge increase in sales value. An avid yachtsman, he organized the Royal Vancouver Yacht Club in 1903 and served as its first commodore. After Walter Graveley's death at eighty-four years of age on June 29, 1939, his ashes were scattered over English Bay. His bequest of Vancouver's earliest competitive sailing award, "The Graveley Trophy,"[8] continues to be prominently displayed at the Royal Vancouver Yacht Club, and Vancouver's Graveley Street is named in his memory.

Lavinia Cortez Fisher and her husband Thomas went on to have five more children. A total of thirteen Fisher children survived infancy—Thomas, Walter, David, Isaac, Lawrence, Lavinia Rosella, Violet, Alfred, Arthur, Rose, Grace, Lillian and Blanche. Thomas Fisher Senior, fondly known as "Old Tom,"[9] and recognized as Vancouver's oldest surviving early pioneer, died on November 18, 1928, at approximately eighty-three years of age. Lavinia died on August 12, 1939, at seventy-eight years of age.

George Cary, often fond of saying that "he was born with a gun in his hand,"[10] continued to be a well-known and respected outdoorsman. He helped to clear a trail into the hills of North Vancouver for real estate agents and was popularly given credit for the logging of a massive North Shore Douglas fir. The tree (a celebrated photograph of which, for a time appeared in the *Guinness Book of World Records*), has generated much controversy over the years. Cary told Major Matthews that he had never cut down a tree of such size, and that the photo was actually a hoax, orchestrated by a group of Vancouver lumbermen at a convention. Debate continues on the true origins of the so-called "Cary Fir,"[11] which some experts have identified as a California redwood. George Cary died on November 4, 1940, at the age of eighty-seven.

Lauchlan Hamilton continued to survey and name Vancouver streets for many months after the fire. Together with federal MP Arthur Wellington Ross, he waged a fierce campaign with the federal government to secure land for the future Stanley Park, as well as Vancouver's first playing field, the Cambie Street Grounds. Widowed prior to his arrival in Vancouver, Hamilton remarried and settled in Winnipeg, where he accepted a position offered to him by Sir William Van Horne as General Land Commissioner of the CPR. Retiring in 1900, Hamilton and his wife Constance purchased a fruit farm in Lorne Park, Ontario, as well as property for a nine-hole golf course. In March of 1938, while celebrating their golden wedding anniversary in Florida, the Hamiltons learned that Lauchlan was to be honoured for his hard work on Vancouver's behalf with the prestigious "Freedom of the City"[12] award. In declining health, Hamilton was unable to make the journey to receive his award in person. Lauchlan Hamilton died in Toronto on February 11, 1941, aged eighty-eight years. A bronze plaque at the southwest corner of Hamilton and Hastings streets commemorates the point from which he began his Vancouver survey in 1885. The many water colour paintings that he completed over the course of his time spent in British Columbia, including Vancouver scenes pre-dating the fire, are preserved in Vancouver City Archives.

Arthur Herring was greatly relieved to hear that the mother of the child he rescued from the Great Fire was found alive and safe in Moodyville. Herring continued in the New Westminster drug business and was active in community and politics, running unsuccessfully for the provincial riding of Westminster in

1890. Frances Herring became a respected novelist and journalist, using many of her husband's recollections from his days as a Royal Engineer for her short stories and books. Frances Herring died of diabetes in 1916. Arthur Herring died on May 29, 1941, at the age of ninety.

While Alexander Strathie began rebuilding the Strathie home at its original Water Street location, Emily spent a week in Victoria, arranging for a $500 loan with a private banking firm. She purchased a new wood stove and brought it back to Vancouver, along with the donation of "a great roast of beef."[13] Work on their house progressed slowly, owing to the tremendous city-wide demand for lumber, and for several nights, the Strathies were forced to sleep under the stars. The stove was set up in the open air on Water Street, where the delicious aroma of roast beef caught the attention of several passersby. As in the days before the fire, generous Emily once again found herself offering comfort food to total strangers. Alexander Strathie died on September 26, 1903, at fifty years of age. Emily Strathie re-married Vancouver City Park Superintendent Mr. George Eldon, and acted as a banker, well before banks were established in Vancouver's early days. She was a lifetime member of the Vancouver Pioneer Association and died on March 17, 1942, at the age of eighty-one.

William Gallagher, much to his sorrow, never learned the fate of the three men from his company who volunteered to help fight the roundhouse clearing fire. He continued to expand his enterprises, developing a successful business in real estate and insurance. In 1896, he was elected to Vancouver city council as alderman. For many years afterward, he remained an outspoken critic of council decisions, ever on the defence against increasing taxation for his long list of clients. He had many vivid recollections of the Great Fire and its aftermath, which Vancouver City

Archivist Major Matthews documented extensively. William Gallagher died on July 11, 1942, at the age of seventy-nine.

Jackson Abray remained with the Vancouver Police Department for four years. He left the force abruptly, fearing for the safety of his family after a prowler's tools were discovered by the window of his home. Returning to the business of hotel proprietorship, he worked for the Cosmopolitan Hotel on Cordova Street, before moving to North Bend, B.C., in 1908. Here he managed the Mountain Hotel, and in later years, a small general store and post office in Boston Bar. Abray's "silver dollar" makeshift police badge can be viewed at the Vancouver Police Museum. Jackson Abray died on April 9, 1944, at the age of eighty-eight.

Annie Ellen Sanders was in attendance when the first gaily decorated CPR passenger train chugged into Vancouver on May 23, 1887. She coaxed a posy of flowers from the engineer and years later donated the pressed and dried petals to Vancouver City Archives. Annie married Stephen Ramage, and together they raised three sons and a daughter. Annie continued with community work throughout her life, helping to organize the first Children's Aid Society in Vancouver and a local branch of the Red Cross during World War I. The last surviving member of the Vancouver Volunteer Fire Department's famed "Coffee Brigade," Annie Ellen Sanders Ramage died on June 5, 1955, at eighty-four years of age.

Hugh Campbell continued serving with Vancouver's fire department as a fireman and electrician until 1899. A talented runner, he competed at many fire-department sporting events. He was employed by contractors Armstrong and Morrison as a foreman, supervising the paving of Vancouver streets. As the last surviving member of Vancouver's original fire brigade, Campbell proudly preserved his red fire vest. He also laid claim to a little-known fact. During Vancouver's first-ever city council meeting on May 10, 1886, young bystander Campbell was recruited to go and find the man being proposed for city clerk. Hugh Campbell, last surviving witness to Vancouver's first city council meeting, died on December 26, 1956, at the age of ninety-two.

Chief August Jack Khatsahlano was the last in a lineage of forty great medicine men in an ancient order of Squamish Nation dancers. The Vancouver neighbourhood name of Kitsilano is derived from the well-known and respected Khatsahlano family. Chief August Jack lived in Squamish for many years, employed as a boom man. He also enjoyed prospecting and carving. Arguably the last surviving individual to have first-hand recollections of the Great Vancouver Fire, Chief Khatsahlano died on June 14, 1967, a month before he would have reached one hundred years of age.

In Memoriam

George Bailey
John Craswell
Mrs. Decker
Albert Fawcett
Mrs. Johnson
Mr. Lambert
Mrs. Nash

And unknown victims of the
Great Vancouver Fire

Afterword: Could it Happen Again?

Could an event like the Great Vancouver Fire ever happen again? Could anything be done to prevent it? These are questions that are tough to answer. Vancouver was by no means unique with its Great Fire experience. Over the course of history, numerous cities throughout the world have suffered their own "Great Fire" through natural or human causes. Among the most famous Great Fires of all time are the ones which occurred in Rome, London, Chicago, San Francisco and Tokyo.

On the night of July 18, 64 AD, a small fire broke out in the southeastern section of Rome's Circus Maximus. Fanned by summer winds, flames spread quickly through the city, consuming tightly packed neighbourhoods of dry wooden structures with ease. The Great Fire of Rome[1] burned for six days, reducing seventy percent of the city to smoldering ash. Rumours spread that Emperor Nero, known to be unhappy with the architecture and layout of Rome, may have been privy to arson. The popular legend that Nero played the fiddle while his city burned is untrue, as the fiddle had not yet been invented.

Thomas Farriner owned a bakeshop in central London's Pudding Lane—a busy establishment with five hearths and a large oven fired up most of the day to bake ship's biscuits and loaves for the poor. Around 1 a.m. on the night of September 2, 1666, Thomas and his family were awoken by the smell of smoke. Wood that had been left to dry alongside the oven had caught fire. The Farriners escaped through a second-storey window and crawled along the gutter to a neighbouring house to sound the alarm. Over the next three days, the Great Fire of London[2]—aided by strong winds and flammable thatched roofs —gutted much of the medieval city.

Chicago's hot, dry summer weather of 1871 had stretched into fall. On the evening of October 8, fire broke out in a barn behind the home of Patrick and Catherine O'Leary. Firemen, exhausted from fighting a large and stubborn fire the day before, misconstrued instructions to go to the O'Leary's and headed for the wrong neighbourhood instead. For two days, the Great Chicago Fire[3] raged through the central business district, levelling most of the city and claiming over three hundred lives. The theory that one of the O'Leary's cows kicked over a lantern, igniting the fire, has never been proven.

In the early morning hours of April 18, 1906, San Francisco was violently shaken to its core by an earthquake estimated (by modern analysis) to have measured 8.25 on the Richter scale. Although the quake itself only lasted about one minute, gas mains were ruptured and wood stoves and furnaces jolted off their foundations, triggering a series of fires which consumed 490 city blocks and laid waste to over 25,000 buildings before burning themselves out three days later. The Great San Francisco Earthquake and Fire of 1906[4] claimed up to seven hundred lives and left over 250,000 people homeless.

The worst fire of them all in terms of human loss occurred on September 1 and 2, 1923, when the Great Tokyo Earthquake and Fire[5] claimed a death toll as high as 142,000. Near high noon, a magnitude 8.3 earthquake centred in Sagami Bay, approximately fifty miles southeast of Tokyo shook the city and neighbouring Yokohama for over five minutes. As it was lunchtime, many people were in the midst of preparing midday meals over gas stoves or charcoal-fuelled hibachis. Japanese homes, largely built of wood with interior paper "shoji" screens, provided ample fuel to quickly increase the spread of the flames. To worsen matters, a typhoon brewing to the north drove dry winds over the area. With water mains broken and no other immediate sources of fire suppressant, fire crews were at a loss to deal with the resulting inferno. A forty-foot tsunami quickly extinguished some of the fires but added many numbers to the death toll. Due to a combination of fire, landslides and tsunami, 60 percent of Tokyo and 90 percent of Yokohama was destroyed.

Through the ages, cities such as New York, Boston, Pittsburgh, New Orleans, Montreal, Toronto, Hong Kong and Shanghai have also experienced "Great Fires." Vancouver's Great Fire is unique among these because it began in semi-wilderness, well outside of the populated area. Once fire reached the outskirts of Vancouver, the classic similarities to other Great Fires began—towering flames ripping through buildings constructed of tinder-dry lumber, driving winds and a fire protection plan that was either non-existent or hopelessly inadequate.

Although the label of "Great Fire" appears to have become too antiquated for modern day use, fires causing major damage to civilized regions are certainly not only the stuff of centuries past.

Southern California residents can attest to years of ongoing battle with wildfires stirred up by the notorious *Santa Anas*—hot, dry, outflow winds originating from the Great Basin and Upper Mohave deserts, which funnel through the state's southerly mountain passes and canyons. In combination with fires generated by lightning strikes, downed power lines, sparks thrown off during machinery operation, carelessly discarded cigarettes or arson, the winds can wreak havoc on civilized areas at any given time of the year. San Diego County's Cedar Fire[6] of late October 2003, started inadvertently by a lost hiker who had lit a small fire to alert rescuers, claimed fifteen lives and destroyed over 4,800 structures.

Populated regions of British Columbia and Alberta have not been immune to catastrophic wildfire in recent years. Kelowna, a thriving, dynamic city strung out along the east shore of Okanagan Lake with a glorious backdrop of orchards, wineries and forested mountains, is a popular locale for retirees and sports enthusiasts. Residents of a south Kelowna neighbourhood may have been initially unperturbed when lightning ignited a fire near tiny Rattlesnake Island—offshore from their community, on August 16, 2003. Over six ensuing days, rogue winds fanned flames of the Okanagan Mountain Park Wildfire[7] over a wide area of south Kelowna, forcing the evacuation of 27,000 residents and destroying 239 homes.

The spring of 2011 was unseasonably hot and dry throughout the province of Alberta. On the afternoon of May 14, fire erupted 15 kilometres east of Slave Lake, a town approximately two hundred kilometres northwest of Edmonton. Within three hours, five hundred hectares of tinder-dry, forested lands were aflame. A massive air assault commenced and volumes of chemical retardant were released to create a fire barrier outside of the town. By the following evening, it was clear that the barrier would be breached. Flames surged, driven by wind gusts of up to one hundred kilometres per hour. At 9:30 p.m. on May 15, a full evacuation order was issued for the 7,000 residents of Slave Lake. The following morning, Slave Lake Mayor Karina Pillay-Kinnee spoke before media cameras and confirmed that 30 percent of the town of Slave Lake was destroyed.[8]

Wildfires have been, and will continue to be a threat to humankind—both locally and globally. Fortunately, while the incidences of fire involving loss of life and property have drastically decreased in modern times, new challenges are on the horizon. There has been much debate about climate change in recent years, with many scientists now in agreement that the upsurge in wildfires, heat waves and damaging storms are direct results of global warming. Could the ramifications of global warming increase Vancouver's vulnerability to wildfire? No one knows for sure, but alarming trends are starting to emerge. Take a walk along a trail or boulevard in any of Vancouver's urban forest settings during a long, hot summer and notice how bone dry your surroundings have become. Fallen twigs break apart with a crisp snap, stream beds lie empty and foliage has become mottled and withered. It is not hard to imagine that a single errant spark in any of Vancouver's urban forests could spell trouble.

Could an accidental or deliberately set fire consume entire business blocks of Vancouver as it did in 1886? It is easy to feel complacent, as many contemporary structures in the Vancouver region are built of concrete and steel, with state-of-the-art fire

An Extreme fire hazard rating in Pacific Spirit Park, October 2012.

barriers and sprinkler systems. Unfortunately, many older structures do not possess all of these modern amenities. In the early morning hours of October 10, 2013, half of a business block on New Westminster's Columbia Street went up in flames due to reasons still under investigation. Although no lives were lost, twelve businesses were destroyed and a well-loved New Westminster heritage building is gone forever.

Now how about the "Big One?" Most Vancouverites are uncomfortably aware that a major earthquake will strike the region sometime in the near or distant future. We have learned of the damage suffered in San Francisco, Tokyo and Yokohama by earthquake and fire. Some risk analysts have pointed out that in the event of a major earthquake, Vancouver's real nemesis would be from fires spawned in the quake's aftermath. Water supply conduits that run under Burrard Inlet could be damaged. Wood pole-mounted transformers could arc and explode, while downed power lines could result in electrical fire. Aging gas pipelines could rupture. The close proximity of older, wood frame houses could easily facilitate the spread of flames. A host of challenges would present themselves for Vancouver and other populated regions of the Pacific Northwest.

The good news is that much progress has been made since 1886. A closer look at the evolution of Vancouver's fire readiness and response plan demonstrates how much we have learned in a relatively short span of time.

Vancouver's original thirty-eight-point Fire By-Law, is now a two hundred-plus page document brimming with regulations on everything from smoke-alarm specifications to the storage of hazardous waste materials. The clearing fires that proved so destructive on June 13, 1886, have long been unlawful in the city. A permit is required for open burning, now restricted to small, contained fires for special occasions, ceremonies and film productions. Forest debris from clearing and storm-damaged sites in urban wilderness regions such as those in Stanley Park and Pacific Spirit Park, are mechanically chipped and sent to city composting facilities, or left to decompose naturally as an integral part of the living forest floor.

George Cary's initial calls for help at the Hole in the Wall Saloon were greeted with skepticism and ridicule. The telephone exchange at Tilley's Books and Stationery had no dedicated line for emergency calls. Fire alarm boxes were non-existent. Cellular networks were years away. Needless to say, we've come a long way in the field of communication since 1886. In the early days of emergency telephone dialing, individual telephone numbers were established to access police, fire or ambulance services, as required. The advent of the three-digit 911 service in 1972 greatly simplified matters. E-Comm 9-1-1[9] is the emergency dispatch center for Metro Vancouver, the Sunshine Coast, Squamish-Lillooet Regional District and Whistler—maintaining a wide area radio communication system that integrates police, fire and ambulance call service. Professional E-Comm operators, highly trained in skills such as data entry, memory recall and decision making, are available twenty-four hours a day, 365 days a year to handle the more than 1.25 million emergency calls received annually. Information provided by callers is relayed by computerized dispatch to the nearest emergency service provider, mobilizing the appropriate response.

When a fire-related emergency call is relayed to any of Vancouver's twenty fire halls, a swift, yet carefully orchestrated series of events unfolds. Upon arrival at the scene of a fire, the commander will quickly determine if additional help is required. Although any fire is potentially serious, fires are broken down into categories, ranging from one-alarm, two-alarm, or three-alarm to higher categories still. The alarm-rating system dates back to earlier times, when crews from fire stations sought help from neighbouring fire stations if they found that a blaze was too difficult to manage. A one-alarm fire is relatively manageable, whereas a three-alarm fire indicates major structural involvement and potential bodily harm. Clearly, Vancouver's Great Fire of 1886 would have registered as a three-alarm in terms of its magnitude, but of course the area did not have three stations to respond.

Despite his good intentions, volunteer fireman Hugh Campbell put himself and many others at risk while wheeling his cart-load of dynamite through Vancouver's crowded streets before a raging wildfire. Today's highly trained Hazmat[10] personnel are on twenty-four hour call with a variety of specialized protocols when an emergency situation involves hazardous solid, liquid or gaseous materials. Vancouver firefighters liaise with the Vancouver Police Bomb Squad to deal with explosives at or near the scene of a fire.

After any immediate hazards have been dealt with, firefighters attack the burning premises with an arsenal of fire-fighting, personal safety and rescue equipment. The nearest beach, creek or well is no longer of consequence, as over 6,200 fire hydrants are strategically situated throughout Vancouver to provide fast and easy access to water. The DFPS (Dedicated Fire Protection System) is a 52-million dollar project consisting of two saltwater pumping stations and a dedicated, earthquake-resistant pipeline to service the downtown core, Kitsilano and Fairview neighbourhoods. Specially-designated blue fire hydrants tap into the DFPS.

Buckets and wet blankets have long been replaced with durable nylon/rubber hoses capable of pumping water at a volume of five hundred pounds per square inch in a minute. Chemical retardants and compressed air foam are carried aboard fire trucks for rapid response to electrical, vehicle or dumpster fires. In the case of a wildland/urban interface fire, cable-suspended helicopter buckets and water bombers may be called upon through the Wildfire Management Branch of the B.C. Ministry of Forests, Lands and Natural Resource Operations.[11]

At the height of the Great Fire, four individuals huddled together in a stable, resigned to a terrible fate. Would modern technology have improved the chances of these unfortunate fire victims? Perhaps it would have. Today's firefighters can be equipped with handheld or helmet-mounted thermal imaging cameras that render infrared radiation as visible light. Thermal imaging cameras enable firefighters to see through smoke, darkness and heat-permeable barriers such as doors or walls to find trapped or unconscious victims. While visibility continues to be a firefighter's nemesis during rescue efforts, technological advancements like thermal imaging are helping to save lives.

Vancouver's 1886 volunteer fire brigade was composed of young pioneering men, many of whom had developed their physical prowess at the end of a bucksaw. Today, it still takes no small degree of stamina to become a firefighter. In addition to various other qualifications, successful applicants must be able to demonstrate superior strength and endurance through the infamous, time-limited Physical Performance Tryout.[12] The Tryout simulates typical scenarios that a firefighter might encounter, such as dragging an 80-kilogram dummy (simulating an unconscious adult victim) to safety, hoisting an 18-kilogram hose roll up 12.3 metres (to a fourth-storey window) and driving a sledgehammer to dislodge a 76-kilogram steel beam (and gain forcible entry to a building.)

Firefighters of today could well be called multi-taskers. Years ago, when wood-burning fireplaces, wood stoves and backyard burning were commonplace in Vancouver, emergency calls for fire suppression occurred much more frequently. Nowadays, Vancouver firefighters are often first responders to any emergency situation. Instead of fighting a fire, they might be called to a car accident, to aid a heart attack victim or even to deliver a baby. Firefighters require extensive first-aid training to supplement the skills of their profession. Strong interpersonal skills, a background in trades or technology, proficiency in a second language, coaching or teaching experience and a commitment to community service are additional attributes of the ideal firefighter.

Just as it was in 1886, wilderness is once again "the back yard" for an increasing number of homeowners. As city population densities increase, families hungry for fresh air and space to play frequently choose to retreat to the scenic suburbs, settling upon forested mountain slopes and rangelands. Homes are often built deep within the forest itself. Here, careless activity can quickly lead to catastrophic results. Statistics maintained by the B.C. Ministry of Forests, Lands and Natural Resource Operations show that although 52 percent of B.C. wildfires are caused by lightning, the remaining 48 percent are caused by human activity. A succession of milder-than-average winters has enabled the mountain-pine beetle to ravage B.C. forests—creating massive regions of bone-dry deadfall, similar to that which proved the ideal catalyst for the Great Vancouver Fire.

In 1886 the citizens of Vancouver received little advance notice that an out-of-control clearing fire was rapidly barreling down upon them. Emergency Information B.C.[13] now maintains constant, province-wide information via radio, television and social media regarding Evacuation Alerts, Orders and Rescinds in the event of any potential threat to life. An "Evacuation Alert" means "Be ready to leave on short notice." An "Evacuation Order" is the most serious. "You are at risk. Leave the area immediately." Local police or RCMP enforce evacuation orders. The "Alert Rescinded" category is undoubtedly a welcome one to any evacuee—"All is currently safe. You can return home. Stay tuned for other possible evacuation orders and alerts."

Mayor Malcolm MacLean recognized early on the importance of designating a gathering point for individuals and families in need of assistance. That duty has now become a coordinated team effort by Emergency Social Services,[14] working under the direction of the B.C. Ministry of Public Services and the Solicitor General. Emergency Social Services is responsible for providing access to food, shelter, clothing, emotional support and family reunification to evacuees. Emergency Social Service reception centres are designated safe areas where evacuees can gather in times of disaster, such as recreation centres, church halls, schools, hotel lobbies or simply tents. Many B.C. hotels display signage notifying guests that they are a designated reception center in the event of an emergency.

While the consequences of a fire that strikes a populated area are disruptive and tragic, in remoter regions, fire can be viewed as an integral part of nature's remarkable predisposition to maintaining ecological balance. Implausible as it may seem, landscapes around the planet—from dense forest to boreal tundra—are dependent upon the periodic intervention of fire for growth and renewal. Forests that are not occasionally subjected to fire become increasingly dense and built up with dead undergrowth. Fire is nature's housekeeper. It "kick-starts" ecological regeneration by creating openings in the forest canopy, which allow sunlight to penetrate and stimulate healthy new growth. Wood ash, rich in minerals like potassium, calcium and magnesium carbonate, produces a strong alkaline reaction that helps to neutralize acidic soil. Indigenous people of centuries past would often set fire to clear tracts of land for agricultural use and to facilitate hunting and travel. Nowadays, under strict governmental regulations, professional forestry crews administer "prescribed burns,"

A contemporary Vancouver fire truck with its motto "Serving with Pride, Since 1886."

fires that have been deliberately set to control deadwood and promote forest regeneration.

Could Vancouver experience another Great Fire? Perhaps more importantly, one should ask, "Are we prepared?" There are numerous ways to access information and learn about fire readiness and safety. Just as it was in 1886, Vancouver is a beautiful and dynamic city with a deeply rooted pioneering spirit—a legacy to carry forward for future generations. As Annie Sanders Ramage put it in her 1948 memoirs, "Grand old Vancouver. I'm glad I came."[15]

AUTHOR'S NOTE

It was not without a certain degree of trepidation that I began to write a book about the Great Vancouver Fire. Would this endeavor be considered too sensational? Too exploitative? At what point does a city's tragedy become a tale of fascination? While attempting to track down the Great Fire experience of Vancouver lifeguard and humanitarian Joe Fortes for my previous book, *Our Friend Joe: The Joe Fortes Story* (Ronsdale Press, 2012), I came across so many long-buried retrospectives—stories of heroism and debauchery, fortitude and folly—that a manuscript not only beckoned, but *demanded* to be compiled.

As is typically the case with research of a historical nature, certain stories become convoluted as eye-witness recollections dim and details are vague in their accuracy or completeness—causing no end of frustration for the writer. Exactly how did William Gallagher arrive back at his downtown office with no apparent knowledge that a massive wildfire was bearing down behind him? Where did John Clough hide his secret stash of blankets? How did the Wah Chong family make their escape? *Vancouver Is Ashes* chronicles, to the best of my capabilities, the events as they occurred on June 13, 1886, and time periods following. Further details to help complete the picture are always welcome, should you have any knowledge of an individual who was "Here before the Fire."

Partial proceeds from *Vancouver Is Ashes* are being donated to the

Vancouver Fire Fighters' Charitable Society. Launched in 1998, this organization raises funds for a wide range of selected provincial charities, such as the B.C. Professional Fire Fighters Burn Fund, B.C. Lung Association and B.C. Children's Hospital. On behalf of the fire fighters, thank you for your support!

ACKNOWLEDGEMENTS

Vancouver Is Ashes would not have been possible without the wonderful involvement and support of many individuals and organizations.

My unbounded appreciation goes to Ronald Hatch of Ronsdale Press, for once again believing in me and facilitating this project with his trademark professionalism, keen eye for detail and remarkable fortitude when being greeted by eighty pounds of love-struck golden retriever. Maria Adams, Deirdre Salisbury and Meagan Dyer relayed my endless volley of emails and draft copies with speed and efficiency. Cover designer Shed Simas employed remarkable talents in giving a new look to an iconic Vancouver photo. Julie Cochrane did her usual exemplary job of page design and layout.

The resources of Vancouver City Archives and Vancouver Public Library Special Collections never cease to amaze. I am most grateful to staff members of these facilities for their expertise in helping me find "the unfindable" on many an occasion. It is always a pleasure to peruse original copies of Vancouver newspapers amidst the historic ambiance of the Provincial Legislative Library. Many thanks to staff members for having requested materials ready-and-waiting during my whirlwind working vacations in Victoria.

Special thanks go to Jillian Povarchook and Wendy Nichols of the Museum of Vancouver for expediting the digitization of Great Fire artifacts for inclusion with this book. As well, thanks go to Kristin

Hardie of the Vancouver Police Museum for photos and pertinent information.

My sisters and colleagues of Native Daughters of B.C. Post #1, along with caretakers Boris and Betty Klavora, have worked long and hard for many years to safeguard the incredible collection of artifacts on display at Old Hastings Mill Store Museum. I am deeply appreciative of their assistance in sharing the history.

The Vancouver Voters List of 1886 is a wonderful resource and I am grateful to members of the B.C. Genealogical Society for the extensive research contained within. Special thanks go to Lorraine Irving for her assistance in this regard. I am also grateful to the following individuals for helping me to fill in the many blanks during manuscript production: John Atkin, Jolene Cumming, Barry Dykes, James Johnstone, Betty Keller, Andy Laycock, Alex Matches, David Mattison, Bruce Macdonald, Latash-Maurice Nahanee, Mike Pearson, Gabe Roder, Drew Snider, Helen Tara, Wendy Warshawski, Bruce Watson, Peter Watson, John Webb and David Williams. If I have inadvertently left anyone out, please consider yourself thanked!

Every year, Vancouver city firefighters host open houses at fire halls throughout the city. I extend my thanks to the crew of Firehall #21 for all the valuable information on modern firefighting techniques.

My heartfelt appreciation goes to Norma Dixon, Dorothy Macey and Nora Schubert of my bi-monthly writing club, whose formidable critiquing skills keep me on my toes, with more than a few light-hearted moments in between. To my husband Doug, along with Hillary, Bobby and Sunny, I love you all and thank you so very much for indulging Mom when she is "in the zone." Lastly, special thanks go to Barbara Rogers, whose extraordinary research skills enhanced the content of this book. Rest in peace, Barbara.

NOTES

SUNDAY, JUNE 13, 1886 (pp. 2–82)

1 James W. Morton, *The Enterprising Mr. Moody, the Bumptious Captain Stamp* (North Vancouver: J.J. Douglas, 1977), 70.
2 Major James Skitt Matthews, *Early Vancouver*, Vol. 4 (Vancouver: City of Vancouver, 2011), 68.
3 Major James Skitt Matthews, "The Great Vancouver Fire of 1886," *Vancouver Historical Journal*, No. 6, September, 1966, 11.
4 Matthews, *Early Vancouver*, Vol. 7, 271.
5 Matthews, "The Great Vancouver Fire of 1886," 11.
6 Matthews, *Early Vancouver*, Vol. 2, 232.
7 R. Monro St. John, "They Were There," *Vancouver Daily Province Magazine Section*, June 16, 1951, 7.
8 R. Monro St. John, "Baptism of Fire," *Vancouver Daily Province Magazine Section*, June 9, 1951, 3.
9 Matthews, *Early Vancouver*, Vol. 2, 233.
10 Matthews, *Early Vancouver*, Vol. 2, 211.
11 Matthews, *Early Vancouver*, Vol. 2, 212.
12 "Chief of Police Stewart's Story," *Vancouver Daily World, 1886 Fire Souvenir Edition*, June 20, 1896, 43.
13 "Story of the Big Fire," *Vancouver Daily World, 1886 Fire Souvenir Edition*, June 20, 1896, 43.
14 "From the Diary of a Doctor," *Vancouver Daily World, 1886 Fire Souvenir Edition*, June 20, 1896, 43.

15 Alex Matches, *Vancouver's Bravest: 120 Years of Firefighting History* (Surrey: Hancock House, 2007), 16.

16 Bessie Lamb, "Origin and Development of Newspapers in Vancouver," MA Thesis, University of British Columbia, September 1942. https://circle.ubc.ca/ (accessed February 2013).

17 Paul Yee, *Saltwater City: An Illustrated History of the Chinese in Vancouver* (Vancouver: Douglas & McIntyre, 2006), 6.

18 Joe Swan, "John Clough: Vancouver's First Pensioner," *West Ender*, September 29, 1983, 5.

19 Matthews, *Early Vancouver*, Vol. 2, 237.

20 R. Monro St. John, "They Were There," *Vancouver Daily Province Magazine Section*, June 16, 1951, 7.

21 Henry Glynne Fiennes Clinton collection, Item AM192, Vancouver City Archives.

22 Matthews, *Early Vancouver*, Vol. 2, 212.

23 "From the Diary of a Doctor," *Vancouver Daily World, 1886 Fire Souvenir Edition*, June 20, 1896, 43.

24 "From the Diary of a Doctor," 43.

25 Arthur M. Herring, "Yesterday was Anniversary of the Big Fire," *Vancouver Sunday Province*, June 14, 1925, 11.

26 Matthews, *Early Vancouver*, Vol. 2, 253.

27 "Saved by the Logging Road," *Vancouver Daily World, 1886 Fire Souvenir Edition*, June 20, 1896, 45.

28 Alex Matches, *Vancouver's Bravest: 120 Years of Firefighting History* (Surrey: Hancock House, 2007), 14.

29 Matthews, *Early Vancouver*, Vol. 1, 105.

30 "History of the Pony Express." http://pnyxpress.tripod.com/history.html (accessed March 2013).

31 Matthews, *Early Vancouver*, Vol. 3, 228.

32 M. Morning Star Doherty, "Community Celebrates History with Canoe Ceremony," www.ammsa.com/publications/ravens-eye (accessed February 2014).

33 "Fire Burned Vancouver 19 Years Ago: Thrilling Story of the Historic Blaze," *Daily Province*, June 10, 1905, 15.
34 "Thrilling Incidents of the Fire Recalled," *Vancouver Daily World, 1886 Fire Souvenir Edition*, June 20, 1896, 13.
35 "Sunday, June 13th, 1886 Was the Date of Vancouver's Big Fire: A Few Reminiscences," *Vancouver Daily World*, June 13, 1892, 3.
36 Matthews, *Early Vancouver*, Vol. 1, 199.
37 "Thomas Dunn's Story," *Vancouver Daily World, 1886 Fire Souvenir Edition*, June 20, 1896, 42.
38 Matthews, *Early Vancouver*, Vol. 3, 236.
39 "Saved by the Logging Road," *Vancouver Daily World, 1886 Fire Souvenir Edition*, June 20, 1896, 45.
40 "From the Diary of a Doctor," *Vancouver Daily World, 1886 Fire Souvenir Edition*, June 20, 1896, 43.
41 "It Came Forth From the Ashes: The Anniversary of the Great Vancouver Fire of 1886," *Vancouver Daily World*, June 13, 1891, 3.
42 "They Dumped the Dynamite," *Vancouver Daily World, 1886 Fire Souvenir Edition*, June 20, 1896, 44.
43 Matthews, *Early Vancouver*, Vol. 3, 236.
44 "Ald. C.A. Coldwell's Experience," *Vancouver Daily World, 1886 Fire Souvenir Edition*, June 20, 1896, 43.
45 Matthews, *Early Vancouver*, Vol. 3, 229.
46 "Fire Burned Vancouver 19 Years Ago: Thrilling Story of the Historic Blaze," *Daily Province*, June 10, 1905, 15.
47 Matthews, *Early Vancouver*, Vol. 4, 111.
48 "Thomas Dunn's Story," *Vancouver Daily World, 1886 Fire Souvenir Edition*, June 20, 1896, 42.
49 Matthews, "The Great Vancouver Fire of 1886," 46.
50 "Appeal for Help," *Ottawa Daily Citizen*, June 15, 1886, 1.
51 "Fled Great Fire on Raft: Mrs. Emily Eldon, Real Pioneer, Dead," *Vancouver Sun*, March 18, 1942, 17.
52 Matthews, *Early Vancouver*, Vol. 2, 212.

53 Matthews, *Early Vancouver*, Vol. 2, 213.
54 Matthews, *Early Vancouver*, Vol. 1, 70.
55 "Vancouver in Ashes: The 'City of Big Hopes' Swept by Flames," *Seattle Times*, June 14, 1886, 4.
56 Matthews, *Early Vancouver*, Vol. 2, 197.

MONDAY, JUNE 14, 1886 (pp. 83–101)

1 Matthews, *Early Vancouver*, Vol. 7, 272.
2 Joe Swan, "Vancouver Police 1886–1887: Growing Pains," *West Ender*, June 2, 1983, 5.
3 Amy I. Kerr, "Took Refuge from the Great Fire on a Makeshift Raft," *Vancouver Province Magazine Section*, May 30, 1936, 5.
4 "Coroner's Inquest," *Vancouver Historical Journal*, No. 3 (January 1960), 49–50.
5 "Coroner's Inquest," 51.
6 Matthews, *Early Vancouver*, Vol. 1, 221.
7 Matthews, *Early Vancouver*, Vol. 2, 253.
8 Matthews, *Early Vancouver*, Vol. 2, 189.
9 "Father Clinton to be Honoured in New Building," *Vancouver Province*, May 5, 1934, 33.
10 "Vancouver in Ashes: The 'City of Big Hopes' Swept by Flames," *Seattle Times*, June 14, 1886, 4.

IN THE DAYS AHEAD (pp. 102–120)

1 "Malcolm Alexander MacLean: The History of Metropolitan Vancouver," www.vancouverhistory.ca/archives_macLean.htm (accessed September 2013).
2 "Mayor Howland requests…" *Toronto World*, June 15, 1886, 1.
3 "Awful Catastrophe, Vancouver City Destroyed by Fire," *Mainland Guardian*, June 16, 1886, 3.
4 Matthews, "The Great Vancouver Fire of 1886," 46.
5 Donald E. Waite, "The Collins Overland Telegraph: The Children

of Fort Langley," www.fortlangley.ca/langley/2etelegraph.html (accessed April 2013).

6 Matthews, "The Great Vancouver Fire of 1886," 47.

7 "Action of the C.P.R. Company," *Daily News*, June 18, 1886, 1.

8 Matthews, *Early Vancouver*, Vol. 5, 142.

9 Matthews, *Early Vancouver*, Vol. 5, 141.

10 "The Fire," *Daily News*, June 17, 1886, 1.

11 "The Fire," *Daily News*, June 17, 1886, 1.

12 "Correction," *Daily News*, June 18, 1886, 1.

13 "Documents Saved," *Daily News*, June 18, 1886, 1.

14 "The Late Mr. Fawcett," *Daily News*, June 18, 1886, 1.

15 "City Notes," *Daily News*, June 18, 1886, 3.

16 "Raised from the Ashes in Three Days," *Vancouver Historical Journal*, No. 6 (September 1966), 67.

17 "City Notes," *Daily News*, June 18, 1886, 2.

18 "Young and Hopeful City of Vancouver Wiped Out by Fire Thirty-eight Years Ago Today," *Vancouver Daily Province*, June 13, 1924, 29.

19 "Fire Burned Vancouver 19 Years Ago: Thrilling Story of the Historic Blaze," *Daily Province*, June 10, 1905, 15.

20 Matthews, *Early Vancouver*, Vol. 1, 203.

21 "Fire Burned Vancouver 19 Years Ago: Thrilling Story of the Historic Blaze," 14.

22 "The Muscular Christians" *Vancouver Daily World, 1886 Fire Souvenir Edition*, June 20, 1896, 13.

23 "Unclaimed goods," the *Daily News*, June 22, 1886, 2.

24 "Photographs Before and After the Fire," *Daily News*, June 26, 1886, 3.

25 "City Notes," *Daily News*, June 18, 1886, 3.

26 Bessie Lamb, "Origin and Development of Newspapers in Vancouver," MA Thesis, University of British Columbia, September 1942. https://circle.ubc.ca/ (accessed February 2013).

27 Derek Pethick, *Vancouver: The Pioneer Years* (Langley: Sunfire Publications, 1984), 167.

IN THE MONTHS THAT FOLLOW (pp. 121–141)

1 "Job Printing," *Vancouver News*, July 23, 1886, 2.
2 "The Fire Bylaw: A Summary of its Clauses," *Vancouver News*, July 24, 1886, 1.
3 "The Relief Fund," *Vancouver News*, July 24, 1886, 1.
4 "Fire! Fire! Testimonial," *Vancouver News*, July 24, 1886, 1.
5 "Echoes of the Streets," *Vancouver News*, July 23, 1886, 4.
6 "Echoes of the Streets," *Vancouver News*, July 27, 1886, 4.
7 "Not Burned but Badly Scorched," *Vancouver News*, July 27, 1886, 4.
8 "Grant and Arkell advertisement," *Vancouver News*, July 23, 1886, 1.
9 "Vancouver's Leviathan," *Vancouver News*, August 4, 1886, 1.
10 "The New Fire Engine," *Vancouver News*, August 2, 1886, 4.
11 "The Auditor's Report," *Vancouver News*, August 10, 1886, 2.
12 Alex Matches, *Vancouver's Bravest: 120 Years of Firefighting History* (Surrey: Hancock House, 2007), 23.
13 "Sir John's Reception," *Vancouver News*, August 14, 1886, 1.
14 "Our Gallant Firemen," *Vancouver News*, August 25, 1886, 1.
15 A. Winifred Lee, "Anniversary of the Fire Recalls Exciting Memories for Pioneer," *Vancouver Daily Province*, May 31, 1941, 11.
16 "The New Fire Engine," *Vancouver News*, August 2, 1886, 4.
17 *Elida Bell Collection*, 1886–1887, Vancouver City Archives.
18 "Christmas Eve," *Vancouver News*, December 25, 1886, 1.

IN THE YEARS TO COME (pp. 142–162)

1 "Clearance Fires," *Daily News-Advertiser*, May 31, 1887, 4.
2 Vancouver Pioneers Association, Vancouver Pioneers Society. List of Survivors, The Scroll of Founders "Here Before the Fire." Box 511-E-2, Vancouver City Archives.
3 Matthews, *Early Vancouver*, Vol. 2, 210.
4 Matthews, *Early Vancouver*, Vol. 7, 271.
5 Matthews, *Early Vancouver*, Vol. 4, 214.

6 *The Great Vancouver Fire*, AM1562-:75-S4, Great fire map—June 12, 1932, Vancouver City Archives.

7 "Major Matthews' letter re Greta Miller's blouse," Museum of Vancouver.

8 "Historic Spot Marked by Pioneers," *Vancouver Sunday Province*, June 14, 1925, 32.

9 "Old Hastings Mill Store Timber Sound after 65 Years," *Campaign re Old Hastings Mill Store to Raise Funds for Moving Building from Harbour to Alma Road, July 1930*. Native Daughters of B.C. Post #1 clipping book, Box 564-D-3 folder 4, Vancouver City Archives.

10 "The Old Mill Store," *Campaign re Old Hastings Mill Store to Raise Funds for Moving Building from Harbour to Alma Road, July 1930*. Native Daughters of B.C. Post #1 clipping book, Box 564-D-3 folder 4, Vancouver City Archives.

11 Donations for Hastings Mill Store relocation and restoration from "Donations of Materials" poster in Old Hastings Mill Store Museum vestibule.

12 Matthews, *Early Vancouver*, Vol. 3, 274.

13 Karen Krangle, "Friday the 13th a Red-Letter Day at Expo," *Vancouver Sun*, June 14, 1986, A1, A3.

WHATEVER BECAME OF . . . ? (pp. 163–178)

1 Peter S.N. Claydon, Valerie Melanson and members of the B.C. Genealogical Society, *Vancouver Voters, 1886: A Biographical Dictionary* (Richmond, B.C.: The B.C. Genealogical Society, 1994), 511.

2 "Former City Clerk Died on Sunday," *Vancouver Daily World*, June 27, 1910, 19.

3 "Alexander, Richard Henry" www.biographi.ca/en (accessed August 2013).

4 "Thomas Dunn and Jonathan Miller: Building Vancouver." http://buildingvancouver.wordpress.com (accessed August 2013).

5 "Forerunner of White Women on Inlet Dies Wrapped in History," *Sun*, June 5, 1916, 12.
6 The History of Metropolitan Vancouver: 1926 Chronology, January 19. www.vancouverhistory.ca/chronology1926.htm (accessed September 2013).
7 Matthews, *Early Vancouver*, Vol. 5, 47.
8 Royal Vancouver Yacht Club. History Committee: The Graveley Trophy (12-2011). http://secure0528.worldsecuresystems.com/announcements/history-committee-12-2011 (accessed February 2014).
9 "Vancouver's Oldest Resident," *Vancouver Sunday Province*, April 19, 1925, 10.
10 Matthews, *Early Vancouver*, Vol. 5, 228.
11 Todd Carney, "*A Disservice to the Douglas Fir: A Fir Tree of the Mind.*" www.cathedralgrove.eu/media/01-1-fir-tree.pdf (accessed July 2013).
12 "L.A. Hamilton, Link With Past, Called by Death in Toronto," *Vancouver Daily Province*, February 11, 1941, 12.
13 Matthews, *Early Vancouver*, Vol. 2, 214.

AFTERWORD: COULD IT HAPPEN AGAIN? (pp. 180–192)

1 "Nero's Rome Burns." www.history.com/this-day-in-history/neros-rome-burns (accessed March 2013).
2 "The Great Fire of London 1666." http://collections.museumoflondon.org.uk/Online/group.aspx?g=group-17548 (accessed March 2013).
3 "The Great Chicago Fire." www.chicagohs.org/history/fire.html (accessed March 2013).
4 "The 1906 San Francisco Earthquake and Fire." www.archives.gov/exhibits/sf-earthquake-and-fire (accessed March 2013).
5 Joshua Hammer, "The Great Japan Earthquake of 1923." www.smithsonianmag.com (accessed March 2013).

6 Nick Carbone, "The Cedar Fire, Southern California, 2003." www.time.com (accessed January 2014).
7 "Okanagan Mountain Park Wildfire." www.kelowna.ca/CM/page129.aspx (accessed March 2013).
8 "Canada's Top Ten Weather Stories for 2011: Slave Lake Burning." http://ec.gc.ca/meteo-weather (accessed March 2013).
9 "Helping to Save Lives and Protect Property." www.ecomm911.ca (accessed July 2013).
10 "BC Hazmat Hazardous Materials Specialists." http://bchazmat.com/ (accessed August 2013).
11 British Columbia Ministry of Forests, Lands and Natural Resource Operations: Wildfire Management Branch. http://bcwildfire.ca/ (accessed August 2013).
12 "Vancouver Fire and Rescue Services Guide for Applicants." http://vancouver.ca/files/cov/Vancouver-Fire-and-Rescue-Services-Guide-for-Applicants.pdf (accessed August 2013).
13 "Emergency Info B.C." www.emergencyinfobc.gov.bc.ca (accessed August 2013).
14 "Emergency Social Services." www.ess.bc.ca/index.htm (accessed August 2013).
15 Matthews, *Early Vancouver*, Vol. 1, 273.

BIBLIOGRAPHY

PRIMARY SOURCES

- **British Columbia Archives**

  British Columbia Archives Vital Events Indexes

- **City of Vancouver Archives**

  Major Matthews Topical Files, Volumes 1–7

  Council minute book 10 May 1886–4 July 1887

  Elida Bell Collection

  Graveley Family Fonds: Legal Papers and Business Correspondence 1884–1901

  Great Vancouver Fire Map, The. AM1562-:75-S4, Vancouver City Archives, June 12, 1932.

  Henry Glynne Fiennes Clinton Collection

  Vancouver Pioneers Association, Vancouver Pioneers Society. List of Survivors, The Scroll of Founders "Here Before the Fire."

- **Museum of Vancouver**

  Major Matthews' letter to Margaret Stevenson

- **Native Daughters of B.C. Post #1**

  Donations of Materials poster

- **Vancouver Police Museum**

Jackson Abray's silver-dollar Vancouver City Police badge

- **Newspapers**

Daily News

Daily Province

Globe

Mainland Guardian

Manitoba Free Press

Montreal Gazette

Nanaimo Free Press

Ottawa Daily Citizen

Seattle Times

Sun

Times Colonist

Toronto World

Vancouver Daily Advertiser

Vancouver Daily News-Advertiser

Vancouver Daily World

Vancouver News

Vancouver Province

Vancouver Sun

Vancouver Sunday Province

Victoria Daily Times

West Ender

PUBLISHED SOURCES

Atkin, John. *Strathcona: Vancouver's First Neighbourhood*. North Vancouver, B.C.: Whitecap Books, 1994.

Barman, Jean. *Stanley Park's Secret*. Madeira Park, B.C.: Harbour Publishing, 2005.

BC Hazmat Hazardous Materials Specialists. Online at http://bchazmat.com/ (accessed August 2013).

Boston Fire Historical Society: "Great Boston Fire of 1872." Online at www.bostonfirehistory.org/firestorygreatfireof1872.html (accessed March 2013).

British Columbia: Emergency Info B.C. Online at www.emergencyinfobc.gov.bc.ca (accessed August 2013).

British Columbia Ministry of Forests, Lands and Natural Resource Operations: Wildfire Management Branch. Online at http://bcwildfire.ca/ (accessed August 2013).

British Columbia Ministry of Justice: Emergency Social Services. Online at www.ess.bc.ca/index.htm (accessed August 2013).

Carbone, Nick: "Top 10 Devastating Wildfires: The Cedar Fire, Southern California, 2003." Online at www.time.com (accessed January 2014).

Carney, Todd. "'A Disservice to the Douglas Fir': A Fir Tree of the Mind." Online at www.cathedralgrove.eu/media/01-1-fir-tree.pdf (accessed July 2013).

City of Kelowna: Okanagan Mountain Park Wildfire. Online at www.kelowna.ca/CM/page129.aspx (accessed March 2013).

City of New Orleans: nofd History. Online at http://www.nola.gov/nofd/about-us/history (accessed March 2013).

City of Vancouver: Vancouver Fire and Rescue Services Guide for Applicants. Online at http://vancouver.ca/files/cov/Vancouver-Fire-

and-Rescue-Services-Guide-for-Applicants.pdf (accessed August 2013).

Claydon, Peter S.N. & Valerie Melanson. *Vancouver Voters, 1886: A Biographical Dictionary*. Richmond, B.C.: B.C. Genealogical Society, 1994.

Davis, Chuck. *The Greater Vancouver Book: An Urban Encyclopedia*. Surrey, B.C.: Linkman Press, 1997.

Dictionary of Canadian Biography: Alexander, Richard Henry. Online at www.biographi.ca/en (accessed August 2013).

Doherty, M. Morning Star. "Community Celebrates History with Canoe Ceremony." Online at www.ammsa.com/publications/ravens-eye (accessed February 2014).

E-Comm 9-1-1: Helping to Save Lives and Protect Property. Online at www.ecomm911.ca (accessed July 2013).

GenDisasters: Montreal, QB Great Fire, June 1852. Online at www3.gendisasters.com/fires/9463/montreal-qb-great-fire-june-1852 (accessed March 2013).

Government of Canada: Environment Canada. Canada's Top Ten Weather Stories for 2011. Slave Lake Burning. Online at http://ec.gc.ca/meteo-weather/ (accessed March 2013).

"Great Fire of 1904, The." Online at www.toronto.ca/archives/fire1.htm (accessed March 2013).

Hamilton, Valerie. *The Schools of Old Vancouver*. Vancouver, B.C.: Renfrew Elementary School, Vancouver School Board, 1986.

Hammer, Joshua. "The Great Japan Earthquake of 1923." Online at www.smithsonianmag.com (accessed March 2013).

Hayes, Derek. *Historical Atlas of Vancouver and the Lower Fraser Valley*. Vancouver, B.C.: Douglas and McIntyre, 2005.

History Files: The Great Chicago Fire. Online at www.chicagohs.org/history/fire.html (accessed March 2013).

History: Nero's Rome Burns. Online at www.history.com/this-day-in-history/neros-rome-burns (accessed March 2013).

"History of the Pony Express." Online at http://pnyxpress.tripod.com/history.html (accessed March 2013).

Johnson, Peter Wilton. *Voyages of Hope: The Saga of the Bride-ships*. Victoria, B.C.: TouchWood Editions, 2002.

Keller, Betty. "Harry Torkington Devine: The Man Behind Vancouver's Historic Photography." *Western Living*, April 1986.

Kerr, Amy I. "Took Refuge from the Great Fire on a Makeshift Raft," *Vancouver Province Magazine Section*, May 30, 1936.

Krangle, Karen. "Friday the 13th a Red-Letter Day at Expo." *Vancouver Sun*, June 14, 1986.

Lamb, Bessie. "Origin and Development of Newspapers in Vancouver." MA Thesis, University of British Columbia, September, 1942. Online at https://circle.ubc.ca (accessed February 2013).

Lee, A. Winifred. "Anniversary of the Fire Recalls Exciting Memories for Pioneer," *Vancouver Daily Province*, May 31, 1941.

Macdonald, Bruce. *Vancouver: A Visual History*. Vancouver, B.C.: Talon, 1992.

Matches, Alex. *Vancouver's Bravest: 120 Years of Firefighting History*. Surrey, B.C.: Hancock House, 2007.

Matthews, Major James Skitt. *Early Vancouver*, Vols. 1–7. Vancouver, B.C.: Self-published, 1932.

Matthews, Major James Skitt. "The Great Vancouver Fire of 1886," *Vancouver Historical Journal*, No. 6, September 1966.

Mattison, David. *Eyes of a City: Early Vancouver Photographers 1868–1900*. Vancouver, B.C.: Vancouver City Archives, 1986.

McNamara, Robert. "New York's Great Fire of 1835." Online at http://history1800s.about.com (accessed March 2013).

McNulty, William J. "Let 'Er Go Gallagher!" *Lewiston Evening Journal*, April 2, 1932. Online at http://news.google.com/newspapers (accessed November 2012).

Morley, Alan. *Vancouver: From Milltown to Metropolis*. Vancouver, B.C.: Mitchell Press, 1961.

Morton, James W. *The Enterprising Mr. Moody, the Bumptious Captain Stamp*. North Vancouver, B.C.: J.J. Douglas, 1977.

Museum of London: The Great Fire of London 1666. Online at http://collections.museumoflondon.org.uk/Online/group.aspx?g=group-17548 (accessed March 2013).

Nichol, Eric. *Vancouver: The Romance of Canadian Cities Series*. Toronto: Doubleday, Canada, 1970.

Papers Past: Great Fire at Shanghai. Online at http://paperspast.natlib.govt.nz/cgi-bin/paperspast (accessed March 2013).

Popular Pittsburgh: Three Tragedies that Changed Pittsburgh. Online at www.popularpittsburgh.com (accessed March 2013).

Regional History from the National Archives: The 1906 San Francisco Earthquake and Fire. Online at www.archives.gov (accessed March 2013).

Royal Vancouver Yacht Club. History Committee: The Graveley Trophy (12-2011) http://secure0528.worldsecuresystems.com/announcements/history-committee-12-2011 (accessed February 2014).

Sandison, James M. *Schools of Old Vancouver: Occasional Paper Number Two*. Vancouver, B.C.: Vancouver Historical Society, 1971.

Sleigh, Daphne. *The Man Who Saved Vancouver: Major James Skitt Matthews*. Surrey, B.C.: Heritage House Publishing Company Limited, 2008.

Smedman, Lisa. *Vancouver: Stories of a City*. Vancouver, B.C.: *Vancouver Courier*, 2008.

Swan, Staff Sergeant G.M. *A Century of Service: The Vancouver Police 1886–1986*. Vancouver, B.C.: Vancouver Historical Society and Centennial Museum, 1986.

"Thomas Dunn and Jonathan Miller." Online at http://buildingvancouver.wordpress.com (accessed August 2013).

Vancouver History: 1926 Chronology, January 19. Online at www.vancouverhistory.ca/chronology1926.htm (accessed September 2013).

Vancouver History: Malcolm Alexander MacLean. Online at www.vancouverhistory.ca/archives_macLean.htm (accessed September 2013).

Waite, Donald E.: "The Collins Overland Telegraph." Online at www.fortlangley.ca/langley/2etelegraph.html (accessed April 2013).

Waite, Donald E. *Vancouver Exposed: A History in Photographs*. Maple Ridge, B.C.: Waite Bird Photos Incorporated, 2010.

Wordie, Jason. "The Great Fire of Hong Kong." Online at www.scmp.com/article/722768/great-fire-hong-kong (accessed March 2013).

Yee, Paul. *Saltwater City: An Illustrated History of the Chinese in Vancouver*. Vancouver, B.C.: Douglas & McIntyre, 2006.

LIST OF ILLUSTRATIONS

PAGE 28 Thomas F. McGuigan. CVA Port P1788. Photo: Stuart Thomson.

PAGE 29 Jonathan Miller. CVA Port P471.

PAGE 30 Andy Linton. CVA Port 268. Photo: George W. Edwards.

PAGE 32 Christina Reid. CVA Port P760. Photo: Wadds Bros.

PAGE 34 Father Henry Fiennes-Clinton. CVA Port P259. Photo: J.D. Hall.

PAGE 36 St. James Church, 1886. VPL 938. Photo: Harry T. Devine.

PAGE 38 Granville, 1885. The large white building left-of-centre is the Sunnyside Hotel. CVA Dist. P14.

PAGE 42 Charles Gardner Johnson. CVA Port P133.

PAGE 43 Jackson Abray. CVA Port N5.

PAGE 46 CPR survey of Hastings Sawmill Site, 1886. Major James Skitt Matthews, *Early Vancouver*, Vol. 3.

PAGE 49 Frank W. Hart. CVA Port P99.1. Photo: J.D. Hall.

PAGE 57 John Joseph Blake. *The World*. Photo: J.D. Hall.

PAGE 61 *Robert Kerr* at anchor in Burrard Inlet, 1886. CVA Wat P30. Photo: J.A. Brock.

PAGE 63 Alderman Charles Coldwell. CVA 371-2915.

PAGE 68 Hastings Mill townsite, July 1886. The two-storey white building (background right) is the Alexander residence. CVA Mi P36.

PAGE 72 Mayor Malcolm MacLean's telegram. CVA AM 54-S16.

PAGE 76 Alexander Strathie's pocket watch. Museum of Vancouver H972.3 111.

PAGE 114 Jackson Abray's silver-dollar Vancouver City Police badge. Vancouver Police Museum.

PAGE 116 Mayor Malcolm MacLean. VPL 3475. Photo: Eldridge Stanton.

PAGE 119 Harry Devine's camera, c. 1886. Museum of Vancouver H990.223.1.

PAGE 124 Vancouver city police force, 1886. Left to right: Const. Jackson Abray, Chief Const. John Stewart, Const. V.W. Haywood, Sgt. John McLaren.

PAGE 127 Cordova Street five weeks after the fire, looking west from Carrall. CVA Str P7.1. Photo: J.A. Brock and Company.

PAGE 131 Vancouver Volunteer Fire Brigade, 1887. Hugh Campbell in middle row, far left. CVA FD P7.

PAGE 133 Vancouver Volunteer Fire Brigade hose reel team outside Fire Hall Number 1, c. 1887. VPL 19792. Photo: Bailey Bros.

PAGE 137 J.W. Horne's real estate "office," 1886. VPL 1099. Photo: Harry T. Devine.

PAGE 139 Fire Hall Number 1 on Water Street, with steam pumper *M.A. MacLean* on right, 1895. CVA FD P41.

PAGE 145 Commemorative ribbon. Museum of Vancouver H982.217.132.

PAGE 148 The Great Vancouver Fire map. CVA 75-54.

PAGE 151 Blouse worn by Greta Miller as she ran from the fire. Museum of Vancouver H979.11.1.

Steam gauge from the *M.A. MacLean* steam pumper. Native Daughters of B.C. Post #1.

ABOUT THE AUTHOR

Lisa Anne Smith was born in Burnaby, B.C., and has been an avid fan of B.C. history for much of her life. She is a longtime education docent at the Museum of Vancouver and is a member of Native Daughters of B.C. Post #1, owners/operators of Old Hastings Mill Store Museum, Vancouver's oldest building. Her published books include *Our Friend Joe: The Joe Fortes Story* (Ronsdale Press, 2012) and *Travels with St. Roch: A Book for Kids* (2001). Lisa lives in Vancouver with her husband, two grown children, and Sunny, the world's least intelligent but most loveable golden retriever.

INDEX

Citations of photographs are in bold

Québec, Canada